The Black Line

Developing the Mission-Planning Software for the SR-71

By John Altson

All rights reserved. No part of this book may be reproduced or transmitted in any form or by any means, electronic or mechanical, including photocopying, recording, or by any information storage or retrieval system without the written permission of the author, except for the inclusion of brief quotations in a review.

ISBN-13: 978-1985693838

ISBN-10: 1985693836

Copyright © 2018 by John Altson

Table of Contents

THE BLACK LINE 1

DEVELOPING THE MISSION-PLANNING SOFTWARE FOR THE SR-71 1

TABLE OF CONTENTS 2

FOREWORD 5

INTRODUCTION BY IAN CAMPBELL 9

THE SR-71 BEGINNINGS, BY IAN CAMPBELL 20

MP^2 – THE AIRBORNE INSTRUMENTS LABORATORY MISSION-PLANNING SOFTWARE 28

The management team (circa 1967) 29

The tools of the trade 30

And now, MP^2... 43

Alpha testing at Edwards Air Force Base 45

Beta Testing at Beale AFB 48

The IBM 7090 at SAC Headquarters 50

Recollections of five of the Airborne Instruments project team 51

An Interview with George Adjami – From Bell Labs to the SR-71, then the LEM, and later real estate 51

Mel Berger – From the SLR Logic to AT&T Labs 56

Joe Cooper – A man wearing many hats 59

Jack Hirschfield – From Republic Aviation to Alarm Systems 66

Joel Raphael – His First Job out of Graduate School 70

A brief history of Airborne Instruments Laboratory (AIL) 79

THE SR-71 MISSION OPERATIONAL FLOW 81

Who planned the missions? 87

What, then, was an "operational mission"? 89

Operational support of the missions 93

Post-mission analysis 97

INTERVIEWS WITH AIR FORCE OPERATIONS 99

Frank Huddleston, Mission Planner 100

Kurt Pfannkuch – After many tours of duty, his SR-71 home at Beale AFB 107

Tony Shelton – Kicking off the post-mission analyses 111

A special story from an SR-71 maintenance expert 115

INTERVIEWS WITH TWO OF THE FLIGHT CREWS 116

Lieutenant Colonel Bredette (BC) Thomas, SR-71 Pilot, 118

Lieutenant Colonel Jay Reid – SR-71 Reconnaissance Systems Officer (RSO) 140

Colonel Gil Bertelson – Surviving an engine fire and the experience of three sunrises on one particular day 146

Colonel Frank Stampf – SR-71 RSO from 1979 until 1983 152

IAN CAMPBELL'S PHOTOGRAPH GALLERY (IAN'S QUOTES) 160

APPENDIX A – A COMPREHENSIVE CREW LIST 182

CONCLUSION 196

AND MANY THANKS TO 197

Foreword

The SR-71 was one of the most amazing aircraft ever built. It was the fastest aircraft ever flown, setting just about every conceivable speed record. To top it all off, it was a very effective spy plane.

Statistics from Wikipedia

Operational highlights for the entire Blackbird family (YF-12, A-12, and SR-71) as of about 1990 included:

3,551 mission sorties flown

17,300 total sorties flown

11,008 mission flight hours

53,490 total flight hours

2,752 hours Mach 3 time (missions)

11,675 hours Mach 3 time (total)

From www.wvi.com

- *New York to London 1 hr 54 min 56.4 sec.*
- *London to Los Angeles 3 hrs 47 min 39 sec.*
- *West Coast to East Coast USA 1 hr 7 min 53.6 sec.*
- *Los Angeles To Washington D.C. 1 hr 4 min 19.8 sec.*
- *St Louis To Cincinnati 8 min 31.9 sec.*
- *Kansas City to Washington D.C. 25 min 58.5 sec.*

32 SR-71s were built. While 12 were lost in various exercisers, not one was lost in operational missions.

Qualitative measures

- Faster than a bullet (really)
- Able to outrun enemy missiles (it did)

Many books have been written about the design of the aircraft, the technical details, and the operational missions. For those wishing to explore the in-depth history of the SR-71, I heartily recommend *Lockheed Blackbird – Beyond the Secret Missions* by Paul Crickmore and *SR-71 Revealed – The Inside Story* by Commander Richard Graham.

There is another yet-unwritten story however – the story of how a small group of talented techies, many of them fresh out of college, banded together to write software that planned the missions and fed the Nortronics navigation system which, in turn, flew the plane. This book is about the team that developed the software and how that software was used for a remarkable twenty-two years. Parenthetically, I was the Airborne Instruments Laboratory Group Leader that managed the mission-planning programming effort.

All the pilots were capable of flying the plane manually but the "auto-pilot" was necessary on the operational missions because of the SR-71's great speed. It would be almost impossible to fly the plane manually and simultaneously manage all of the sophisticated instrumentation.

The Black Line was written for a general audience and SR-71 enthusiasts.

This book would not be possible without the contributions of many people:
- Ian Campbell, a military historian from Australia who wrote the first sections: an introduction and an overview of the SR-71 beginnings.

- My Airborne Instruments Laboratory team members, doing their best to recall the project from fifty years ago.
- Several Air Force pilots and RSOs, recalling their flight experiences and the use of the A.I.L. software.
- Several Air Force SR-71 maintenance personnel, detailing the evolution of the software.

What you will read in the ensuing chapters are:
- Ian's gripping story about the SR-71 press releases from the LBJ archives
- Ian's brief SR-71 history
- The development of the mission-planning software, with team interviews
- The SR-71 (spy) mission operations flow of command
- Interviews with Air Force operations personnel
- Interviews with two noteworthy flight crews. One pilot, Gil Bertelson, passed away in December of 2017
- Ian's incredible photo gallery, with his own flight diagrams
- A comprehensive flight crew list

The SR-71 flew many missions and, as you might expect, there were quite a few stories from those many missions. A few such stories are mentioned later in this book. Here is an excerpt from just one of those stories:

"The Emergency, by Lieutenant Colonel BS Thomas"

"After transiting Murmansk and while hooked up with the tankers during the fourth refueling, I saw a flicker of the Master Warning light.

Jay Reid announced the light at the same time that I saw it in my peripheral vision. I disconnected from the tanker's boom and maneuvered back to the pre-contact position. Our indication was the illumination of the left-engine oil supply low-quantity red warning light. From our training and experience with the aircraft emergency checklists, we knew immediately that this required that we "land as soon as possible."

The mandate to land immediately was borne out of long experience with malfunctions: the type that would crash the airplane if another single-system failed.

The natural tendency for military aircrews is to complete the mission if humanly possible. To counter this inclination, the Wing Commander had designated certain emergencies sufficiently critical to require immediate landing. This was one of those emergencies."

Enjoy the book!

John Altson

Introduction by Ian Campbell

Even at this remove, more than twenty-five years after its retirement, it's difficult to describe the Habu without repetition or hyperbole. Or both.

What we've attempted to do in this work is look at the aircraft and the program from a hitherto entirely unexplored direction: the mission-planning and operations software developed by Airborne Instruments Laboratory, with which John Altson was involved. Not only has this never been touched on before, but he offers an unparalleled insight into a very famous aircraft from a contractor's point of view. John was involved from 1964 to 1967, crucial years in the development of the aircraft, introduction into operations, and the missions it undertook.

Workers at Lockheed's Aircraft Design Project (ADP) said Kelly Johnson, the brains behind Skunk Works, could almost "see" air moving. He could look at a design or model and know roughly whether or not it would be efficient. Not only that, but he could estimate its drag-coefficient within a few percent of what company aerodynamicists would spend nights in a wind tunnel and days calculating.

Today, Kelly's leadership and management style would be defined as abrasive, hectoring, and occasionally abusive or bullying. But he was one of the last of the classic auteur designers. Such men knew every aspect of aircraft design and production. Kelly came from a background of well-rounded men able to turn their hands to anything, physical or theoretical. He rose from manual and skilled laborer to engineering college, to university, to a revived, redirected fledgling Lockheed as a tool designer in 1933. He genuinely worked his way up in a world and a system that could never happen now. Changes in lifestyle and work style, different approaches to education and training, CADCAM – all have irrevocably altered how life prepares us.

It's safe to say that during 1940-1970, Johnson's designs kept Lockheed solvent. Design changes to the Electra airliner and to the Hudson patrol-bomber pushed every limit of materials and theory at the leading-edge of aerospace with P-38, P-80, F-104, U-2, A-12 and SR-71, even non-Skunk Works projects such as the immortal C-130 Hercules. Kelly occupied an apogee with a tiny group of equals – Storms, Sukhoi, Tupolev, Kartveli. Lockheed-Georgia's Evergreen Herky Bird was built on Skunk Works principles. C-141 and C-5A used ADP design experience.

Lockheed might well have gone bankrupt from corruption, overstretch, and complacency, had foreign sales (F-104), maintenance (C-130, SR-71, U-2) and black funding (A-12, M/D-21, drones) not kept the lights on. It was not bad design that almost led to the company's fall.

Part of the permanent attraction was how secret and mysterious these aircraft were. It's easy to forget, today, when so much is known and even more is available online. Even though it was not a black program as all its forerunners were, the Habu was heavily classified. What was known, or was at least "common knowledge," was limited, often wrong, and exaggerated – sometimes deliberately.

What was known was also much less than what we remember. Though open and acknowledged from September 1964, there was precious little official information beyond very basic stats.

Numbers, deployment, systems, and capabilities were briefly summarized; a few news stories here and there pertaining to events, accidents and losses, but little else. Compared to fighters and bombers, the SR-71 remained in shadow, beyond a few stats and the same few photos, reprinted endlessly. These were always official, carefully vetted, and general. They were selected, like F-117 and B-2 photos a generation later, to avoid any close-ups or details, to conceal anything subtle about the plan, form, shaping, size of fleet, and activities – and, above all: signature-reduction. Therefore, photos of SR-71A 61-7961, a notorious hangar-queen, with infamous 'dark indigo blue' color prints sent to magazines and books are still used today, though they date from the mid-1960s. One key indication is the early silver pressure suits. Another is the full white insignia with red-white-blue national roundels, developed at such expense to survive heat soaking. Neither was used after 1981-83, nor were the tiny dents indicating ECM in the nose behind the pilot probe. If you can find a photo of an SR-71 with ECM bulges and stars and bars, you know it dates from 1979 and 1983, after ECM upgrades and before the last of the fleet was repainted at depot maintenance.

Stealth might have been a given from the 19th century in submarines, but it was wholly unknown for aircraft before 1980. Still less was the idea that an aircraft could be built to reduce something – radar-return – that almost everyone knew nothing about.

Even today, people assume radar is like sonar. It pings, or at least makes a noise – and works underwater. Still less, did anyone in 1964 realize anyone could design an aircraft to be smaller on radar without compromising aerodynamics, let alone mach-3 aerodynamics?

Compared to the seemingly endless stream of information and mis-information on anything and everything these days, it was a totally different world.

Unseen outside the US (except at Kadena Air Base in Okinawa) from 1968 until 1974, Habu retained an unparalleled mystique. In September 1974, not only did the SR-71 set official trans-Atlantic records: New York to London and London to Los Angeles, it appeared, like a returning Apollo, from behind SAC's veil of secrecy for the first time. Photos and TV were made, interviews to press given, secrecy (a bit) reduced. Now ten years old and with eight years' continuous Vietnam service behind it, there was still almost nothing the public could be told about it. Beyond the obvious.

From the beginning, disinformation and misinformation were part of the show. Developed and deployed in absolute secrecy from 1959 to 1963, CIA's revolutionary A-12 mach-3+ titanium photoreconnaissance platform stayed that way past 1964, 1968, and 1974 – until 1982. Despite being the size of a B727 tri-jet regional airliner, Kelly's "Archangel" didn't exist. Only when it needed to move into high-Mach and high-altitude test and development outside the few hundred thousand square miles of airspace reserved for it, did anyone even think about a cover story. Sooner or later some Air Force or airline pilot – or passenger – was going to see something large and shiny above, below, or beside them and tell people – other pilots, family, newspapers. Someone might take some photographs.

This was, after all, still the age of UFO sightings and B-movie sci-fi. People had been seeing shiny things in the sky for years. Sometimes, after 1955, they were CIA and then Air Force U-2s at 70,000 ft. Sometimes, they weren't. Exactly how – and how much – to release would occupy political and intelligence committees for over two years.

Smarting from Republican claims of being soft on Commies and Defense, President Lyndon Baines Johnson decided to move on the issue and surface a black program. It was clear Oxcart was going to be sighted more often, and by people who knew what they were looking at. Rumors abounded in the trade press. Something would have to be done to protect Oxcart, solve LBJ's quandary, and end a long contentious debate.

Still smarting from repeated claims that he wasn't doing enough in Vietnam or elsewhere, in September 1964 LBJ moved to demonstrate just how far U.S. superiority had progressed.

The R-12 / RB-12 / and RS-71 program due to fly in December would be announced early enough for PR and political pressure to take effect before the vital general election in November.

You will find the following (now declassified) memoranda of interest.

Yes, this is unclassified. And none of it was my work. Fellow Habu devotee Alan Johnson requested it from the LBJ Library to resolve ongoing debate (at three different Facebook pages!) on whether the RB-12 / R-12 / RS-71 / SR-71 designation was LBJ's mere slip of the tongue (covered up by White House PR editing), or the result of a longer-term campaign of misinformation and disinformation intended to disguise the CIA origins and cost of the A-12 / YF-12A, and how long it had been in development.

Even LBJ's 1964 press conferences are on YouTube for our edification.

LBJ YF-12-A 'A-11' Announced 29 Feb 1964
https://www.youtube.com/watch?v=ZGxLbqnDKJU&fref=gc
LBJ Press Conf. 24 Jul 1964
https://www.youtube.com/watch?v=_GLxcw_hObY&fref=gc

I think, more than anything else, it indicates not only the importance of accepting reasonable claims from eyewitnesses, but also trusting that primary sources are preserved and will, generally, establish what actually happened.

THE WHITE HOUSE
WASHINGTON

~~TOP SECRET~~

January 18, 1963

Dear Jim:

Pat Coyne tells me that you are convening a number of members of the Board to consider alternative plans for dealing with the Air Force development of a plane that has interesting relatives. In this connection, I thought I ought to tell you that there is very considerable concern among technical people in the Government -- Harold Brown, Jerry Wiesner, Spurgeon Keeny and others -- about the new plan which Pat brought up to you on his last trip. This concern is that if McNamara makes a presentation along these lines to the thirty-odd members of two major committees and two major subcommittees, even if the presentation is Top Secret, there will get to be gossip about a new plane of just the sort that will lead suspicious people toward its relative. This is the more so because the other development is so far along that there will be plenty of unexplained events in the airways and on the airfields of the United States in the next few months. So these people are inclined to the view that this middle course has the worst features of both a black and a white operation, in a time in which we are not dealing with a virgin field as far as the possible ideas of technical reports and journals are concerned.

For this reason, Harold Brown and others in the Defense Department are now playing with the idea of going back to the original McNamara presentation, by giving the new plane a different and bogus set of relatives in the field of long-range interception. This they think would work for a while and be much less suspicious than a gradually spreading bit of Top Secret briefing.

My own preference from the intelligence view is for a completely black operation in all of these vehicles, for as long as it lasts, although I am not really very hopeful that the period of grace ahead is very large. (On this I really think I probably disagree with Din Land and some others who, I fear, are assuming too readily that an

~~TOP SECRET~~

- 2 -

~~TOP SECRET~~

early history of success can be repeated in a different environment.)

These are random and hasty comments, but in talking with Pat I thought they might be of some relevance to your coming discussions.

Since Bob McNamara's budget presentation comes about Wednesday of next week, time is getting short and I know the President will need to have your opinions by the first of the week at the latest, if he is to be able to weigh them effectively.

Sincerely,

McGeorge Bundy

The Honorable James R. Killian, Jr.
Chairman of the Corporation
Massachusetts Institute of Technology
Cambridge 39, Massachusetts

~~TOP SECRET~~

STENOTYPE TRANSCRIPT
OF
PRESS CONFERENCE

Press Office Files, Backup Material 7/24/64, box 69
LBJ LIBRARY

PRESS STENOTYPISTS ASSOCIATION

Specialized Press Reporting

Washington, D.C.

306 Ninth St., N.W.
NA. 8-3406

917 G Street, N.W.
NA. 8-4266

PRESS CONFERENCE NO. 23

of the

PRESIDENT OF THE UNITED STATES

3:30 P.M. (EDST)
Friday,
July 24, 1964

Auditorium,
State Department,
Washington, D. C.

THE PRESIDENT: Good afternoon, ladies and gentlemen. I would like to announce the successful development of a major new strategic manned aircraft system, which will be employed by the Strategic Air Command. This system em-loyes the new RS-71 aircraft, and provides a long range advanced strategic reconnaissance plane for military use, capable of worldwide reconnaissance for military operations. The Joint Chiefs of Staff, when reviewing the RS-70, emphasized the importance of the strategic reconnaissance mission. The RS-71 aircraft reconnaissance system is the most advanced in the world. The aircraft will fly at more than three times the speed of sound. It will operate at altitudes in excess of 80,000 feet. It will use the most advanced observation equipment of all kinds in the world. The aircraft will provide the strategic forces of the United States with an outstanding long range reconnaissance capability.

The system will be used during periods of military hostilities and in other situations in which the United States military forces may be confronting foreign military forces.

The RS-71 uses the same J58 engine as the experimental interceptor previously announced, but it is substantially heavier and it has a longer range. The considerably heavier gross weight permits it to accomodate the multiple reconnaissance sensors needed by the Strategic Air Command to accomplish their strategic reconnaissance mission in a military environment.

This billion dollar program was initiated in February of 1963. The first operational aircraft will begin flight

Page 1 - #23

This Copy For _____

PRESIDENT LYNDON B. JOHNSON'S NEWS CONFERENCE #23

Held in the State Department Auditorium
23rd and C Streets, N.W.
Washington, D. C.

July 24, 1964
At 3:30 P.M. EDT (Friday)

In Attendance: 315

Official White House Transcript

THE PRESIDENT: Good afternoon, ladies and gentlemen.

I would like to announce the successful development of a major new strategic manned aircraft system, which will be employed by the Strategic Air Command. This system employs the new SR-71 aircraft, and provides a long-range, advanced strategic reconnaissance plane for military use, capable of worldwide reconnaissance for military operations.

The Joint Chiefs of Staff, when reviewing the RS-70, emphasized the importance of the strategic reconnaissance mission. The SR-71 aircraft reconnaissance system is the most advanced in the world. The aircraft will fly at more than three times the speed of sound. It will operate at altitudes in excess of 80,000 feet. It will use the most advanced observation equipment of all kinds in the world. The aircraft will provide the strategic forces of the United States with an outstanding long-range reconnaissance capability. The system will be used during periods of military hostilities and in other situations in which the United States military forces may be confronting foreign military forces.

The SR-71 uses the same J-58 engine as the experimental interceptor previously announced, but it is substantially heavier and it has a longer range. The considerably heavier gross weight permits it to accommodate the multiple reconnaissance sensors needed by the Strategic Air Command to accomplish their strategic reconnaissance mission in a military environment.

Press Office Files, Backup
Material 7/24/64, box 69
LBJ LIBRARY

PRESS RELEASE

I would like to announce the successful development of a major new strategic manned aircraft system which will be employed by the Strategic Air Command. This system employs the new SR 71 aircraft, and provides a long range advanced strategic reconnaissance system for military use, capable of world-wide reconnaissance of military operations. The Joint Chiefs of Staff, when reviewing the RS 70, emphasized the importance of the strategic reconnaissance mission.

The SR 71 aircraft reconnaissance system is the most advanced in the world. The aircraft will fly at more than three times the speed of sound. It will operate at altitudes in excess of 80,000 feet. It will use the most advanced observation equipment of all kinds in the world. The aircraft will provide the strategic forces of the United States with an outstanding long-range reconnaissance capability. The system will be used during periods of military hostilities, and in other situations in which United States military forces may be confronting foreign military forces.

The SR 71 uses the same J-58 engine as the experimental interceptor previously announced, but it is substantially heavier and has a longer range. The considerably heavier gross weight permits it to accommodate the multiple reconnaissance sensors needed by the Strategic Air Command to accomplish the strategic reconnaissance mission in a military environment.

This billion dollar program was initiated in February 1963. The first operational aircraft will begin flight testing in early 1965 and deployment of production units to the Strategic Air Command will begin shortly thereafter.

Appropriate members of Congress have been kept fully informed on the nature of, and progress in, this aircraft program.

Further information on this major advanced aircraft system will be released from time to time at the appropriate military secrecy classification levels.

Highly classified — all defense info appropriate today — Will background further today

The SR-71 beginnings, by Ian Campbell

It was unfortunate no-one knew then what a challenge M3+ flight was going to be, let alone providing passenger comfort.

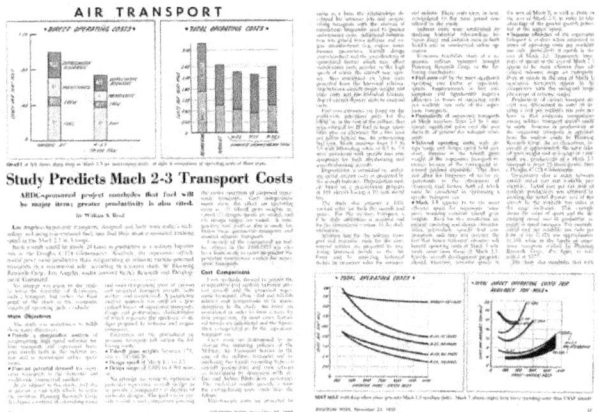

Even for an optimistic decade on the verge of the Space Race, this was unrealistic. NAA's XF-108 had been cancelled in September 1959, and the XB-70 schedule was already slipping. In Europe, only rocket-jet experimental aircraft were exploring this region, as NASA had already moved beyond X-3 and X-7 to higher speeds and altitudes in the X-15, but not for sustained periods. And not in designs that could take off and land under their own power, let alone operate routinely.

Lockheed, the only Westerner to get a M3+ design to fly (which wasn't a jet/rocket-propelled X-Plane), was only just beginning to discover at the Skunk Works what the thermal and materials challenges of M3+ were.

ADP had barely started cutting test components, or flown anything. People forget how many problems ADP, etc. had to confront and overcome from 1957 to 1962. Once flying, it was three more years F-Test before it was flying at design speed and altitude to operational range. As a working intelligence asset, Oxcart had to fly for 2,500 miles and two hours, decelerate, descend; AAR, accelerate, ascend; do another leg, and then decelerate; AAR accelerate and return to base. All without regularly losing an engine, fuel, navigation, air conditioning or have a camera failing. If the cockpit air-conditioning failed, the pilot could bake a cake in his lap.

Everything on the aircraft had to be designed and built from scratch, not just the engines, inlets and exhaust, but fuel / fuel system, hydraulics, using the fuel as heat sink, stopping tires failing from heat-stress.

Designing a one-ton five-foot-tall 18"sq negative fine-grain film-camera to operate not just at 80,000 feet and −90C, but at 3.2M, where the temperature on the outside of the aperture glass was 300-500F, wasn't just a matter of adapting that massive camera from the U-2. Hycon, Kodak, and Perkin-Elmer cameras had to be re-engineered for higher altitude, larger temperature cycles (as U-2s went high and stayed cold) both hot and cold, and for greater vibration and damping. Navigation was now a serious issue, as the platform moved a mile every two seconds, not a mile every eight seconds − 400kts (470 mph), where pilots had time for sun sights and manual astro-nav. No INS existed sensitive enough to handle such speed for such duration, compared to ICBMs or airliners.

Yes, it's true that out of this program the more mature SR-71 emerged, with simplified construction (now ADP had learned how to build a M3+ spy plane, at eye-watering cost), so that a single strong component was now cast, machined and installed, instead of using the former process and being assembled out of six to eight smaller components. The only way all this was possible, even without the compressed timeframe (1959-1962 for actual design, development and construction), was with an ocean of secret money. F-4 Phantoms were $8-9 million and even the very-expensive RA-5Cs (with thin films of gold-plating on both engine bays) were $13 million each.

At the same time, the individual cost of 15 A-12s amortized over the total R & D cost of the project was something approaching $254 million each. That at least included engines and sensors. $250 million is a lot for a jet today, unless it's a 200-seat airliner. Fifty-five years ago, it would have caused heart attacks in DC, let alone in suburban America.

Two billion dollars ($2,000M) was spent by Pratt and Whitney to develop the J-58 from a single-use cruise-missile engine into a fully bypassed multi-inlet turbojet with afterburner capable of withstanding M3+ flight for four to eight hours or longer. An entire factory was carved out of the Florida Everglades.

None of this made OXCART more attractive to its paymasters or customers, but it did pay off for SAC. Like the U-2 before it, LeMay let the CIA pay for design and construction and then ensured he got a mature White-World platform for the next decade – and, as it turned out, – the next quarter-century.

Hundreds of men worked twelve to fourteen-hour days, five or six days a week for years on end, supported by tens of thousands of contractors, none of whom were allowed to know what they were building, even while they were inventing parts of the most challenging aerospace project ever built.

Compared to the heat-soak and structure problems of the A-12, the X-15 was child's play: ten minutes at M7 was barely an issue. Even Apollo heat shields presented fewer issues: for one, they only withstood extreme temps for a limited time (15 minutes from entry to splashdown). Moreover, they only needed to work once. They were not designed or expected to operate routinely, several days a week, for years at a time.

No one expected the A-12 to be in use for six years, or thirty SR-71s for twenty-five years. Yes, the titanium annealed from the heat of every flight and reputedly became stronger, but that didn't apply to other parts such as engines, elevator actuators, and elevator cables (the latter made from Elgiloy, an alloy specifically engineered for watch springs, not aeronautics).

One A-12 baked and wrote off its entire wiring loom early in flying to maximum Mach in 1963-1964, simply by getting hotter than predicted flying faster, and the aircraft was nearly lost. Another had its engine-ejector flaps wired shut for a ground test. They were forgotten before the next hi-Mach run and the aircraft "took off on its incredulous pilot" after air refueling "like a scorched dog", and he, too, only just made it back. In January 1967, with cancellation staring the program in the face, a fleet-wide modification to ejection seats was made. The modification followed an SR-71 accident fatality that led to the horrible death of an OXCART pilot. After his A-12 ran out of fuel (faulty gauge and inadequate test-plumbing on the ground), and he made a safe exit, the seat-separation mechanism failed, and he rode the seat into the ground.

Even once OXCART was declared operational in 1965, it was over two years before it was finally used, when on the verge of being cancelled outright. It was now in competition with the (White World) SR-71, despite the fact the latter aircraft had smaller cameras with lower resolution. Even a competitive fly-off between both aircraft and systems only highlighted this issue, instead of solving it.

Mis-and-disinformation were just as present as they had been with the U-2's infamous "NASA Weather Research Plane" cover story, but much more subtle and skilled. Programs were hidden in plain sight, routinely renamed or mislabeled, officially assigned to other Federal agencies, funds and contractors obscured – especially from each other. Normal information available on ordinary aircraft was simulated and released so that everything looked and sounded much smaller and routine.

Had the public known that $2 billion had been spent just on J-58 engine development and associated inlet-exhaust R & D, it might have been a different matter – especially when the CIA received twenty U-2s for $22 million in eighteen months. Oxcart posed challenges where the orders of magnitude were more complicated.

Even in 1971, NASA's "YF-12C", of course, wasn't. There was an entire pointless back-story covering why NASA was using a prototype, cutting-edge reconnaissance aircraft for SST research (admittedly a flight-test bird, seven years after First Flight). Not only was "YF-12C" spurious, but also 1960-06937 had been an actual operational A-12, not just a consecutive serial after the genuine YF-12As `934, `935 and `936.

On SST Technology

We now know there was a great deal more information in the public arena than previously assumed, 1962-67. We can point to specific 1966 _Aviation Week_ articles where both Lockheed and Pratt & Whitney were showing their advances, not specifically re Oxcart and Senior Crown but more generally and under the guise of America's SST program.

The following images clearly show work for the CL823 - L2000 based obviously on work on the J58 nacelle, shock cone and bypass. That was the most advanced, most powerful working supersonic inlet and turbojet in the world. Despite all its huge size and enormous thrust, Pratt and Whitney's J93 for the XB-70 was already relegated to mach 2.5.

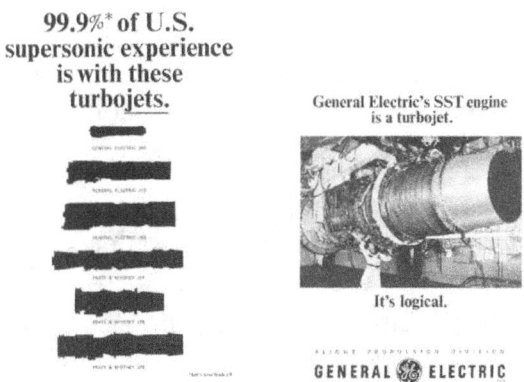

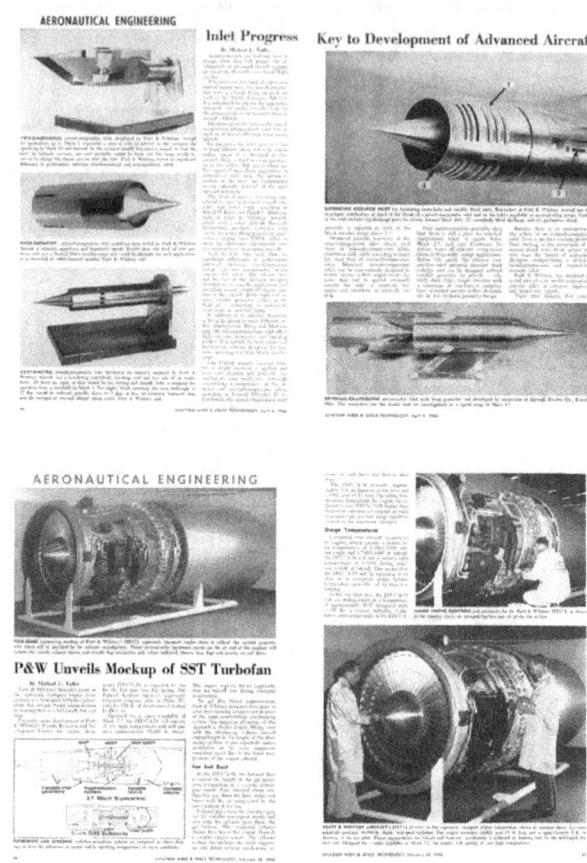

With the loss of AV2 [62-0207] in June 1966 and the critical honeycomb/skin bonding issues of AV1, the entire Valkyrie program was limited to mach 2.5. This was of no moment to the US SST, which was intended for mach 2.4.

It must have been tremendously galling for the National Aeronautical Association (NAA), first stymied losing the XF-108 for ADCom, then XB-70 changed into a mere research program by government strategy changes & funding cutbacks. Now their revolutionary and prestigious achievement was reduced to a single surviving aircraft, in the knowledge Lockheed or Boeing were going to win the future SST contract.

Worse still, that NAA had been outmaneuvered and outperformed by a small rival division operating in the "black world", enjoying enormous secret funding and support they'd been denied for years.

Not that Lockheed generally or ADP in particular felt tremendous about the future. Three times the demonstrated F-12 capability was voted funding for a fleet of 90 continental interceptors, and three times McNamara's Whiz Kids had vetoed it.

Oxcart remained black; the small Air Force SR-71 fleet would replace it, but no more. In the future would come an equally galling ignominy. McNamara would demand the progeny be killed as soon as it was fledged: all the tooling, jigs and dies were to be destroyed. Sold as scrap for 7.5 cents a pound.

Further mach.3 research was ended, beyond small NASA programs around the edges. No more hardware, no next-generation progress, and no capability to meet a future need. While the threat, especially in range, of the Ty22M2 ["Backfire"] would be exaggerated in the 1970s & 1980s, no one knew this in 1968, or 1973, or 1980.

Just as no one knew the Ty160 would eventually produce intercontinental supersonic capability, that no follow-on from Ty144 would yield results, or that the entire USSR would collapse.

The indications were there (even in 1970), but no one bar a few CIA and freelance analysts accepted them.

MP² – The Airborne Instruments Laboratory mission-planning software

How's your memory going back fifty plus years? My story, and the stories contributed by my co-workers, is the result of a true memory test.

It was early in 1965. I was, at the time, a Senior Programmer working in the Applied Mathematics Division at Brookhaven National Laboratory in Upton, Long Island. I had been there for almost three years and was "getting antsy," as the saying goes. Brookhaven was a wonderful place to work. Five weeks of vacation each year, softball in the afternoons, and not too much pressure. Why in the world would I want to leave? Maybe I was not challenged enough? Maybe it was for career advancement? In any event, the Long Island newspaper, *Newsday*, ran a full-page ad for programming jobs at all levels at Airborne Instruments Laboratory in Deer Park, Long Island.

I went for an interview and was hired on the spot. I had no idea of the nature of the project for which I was hired. When I reported for work, I was dispatched to a cubicle and told that I needed to prepare an application for Top-Secret Clearance. I was also told that I could not learn about my ultimate project work until my clearance came through.

I filled out the application for Top-Secret Clearance; it required a tremendous amount of homework, as I remember. Where had I lived at every stage of my life? What were my parents' backgrounds? My grandparents' backgrounds? Sibling histories? I wish I had saved a copy of the application.

So, I waited. Those in the adjacent cubicles waited. We all waited for our clearances to get processed. While we waited, my manager, John Boccio, gave me mathematical subroutines to code in the FORTRAN language and test. They were ridiculously easy. I knocked them out in minutes and wrote "inverse subroutines" to check the logic. John Boccio was impressed.

Not all of us were patient. The young man in the adjacent cubicle gave up and took a job at Boeing. The rest of us stayed for about six weeks until the clearances came through.

The management team (circa 1967)

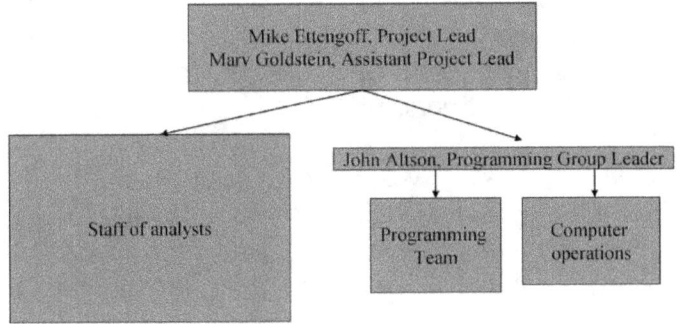

Functional Organizational Chart

Mike Ettengoff and Marv Goldstein (both now passed away) ran the project effectively. Mike led the contract negotiations while Marv was more involved day-to-day. The profit and loss responsibility rested on the shoulders of Mike and Marv (this was a "cost-plus" Air Force contract, so there was no concept of "loss").

I soon reported directly to Marv and, I have to say, he was an excellent manager, good to me personally. Marv also managed the team of systems analysts that, in turn, provided the logic for the programming team.

The tools of the trade

In 1965, the computational tools were primitive by today's standards. "Bearskins and Knives," one might say. The "state-of-the-art" computer on which we worked was the Control Data 3200. It had 32,768, 24-bit words and processed about one million instructions per second. It was classified as a "supercomputer" in 1965. I now have an Apple iPhone 6 with one gigabyte of memory. The iPhone 6 therefore has roughly 10,000 times the memory of the CDC 3200. The iPhone 6 processes about 3.36 billion instructions per second and has over 3,000 times the processing power of the CDC 3200. What a dose of perspective!

In 1965 there were no disk drives, no laser printers, and no PCs. The "inputs" were "IBM" punched cards, prepared on an 026 keypunch. The outputs were (binary) punch cards, and printouts to a "high-speed" 600-line-per-minute line printer. Mass storage was done on magnetic tapes.

The CDC 3200

CONTROL DATA® 3200 Computer System / Real Time Applications

Here, from Wikipedia, is a detailed description of the Control Data 3200:

"The CDC 3000 series computers from Control Data Corporation were mid-1960s follow-ons to the CDC 1604 and CDC 924 systems. Over time, a range of machines was produced - divided into the upper-3000 series and the lower-3000 series. CDC phased out production of the 3000 series in the early 1970s. The 3000 series were the cash cows of Control Data during the 1960s; sales of these machines funded the company while the 6000 series was designed.

The upper 3000 series used a 48-bit word size. The first machine to be produced was the CDC 3600; first delivered in June 1963. First deliveries of the CDC 3400 and CDC 3800 were in December 1965. These machines were designed for scientific computing applications, however were overshadowed by the 60-bit CDC 6000 series machines when the CDC 6600 was introduced in December 1964 and delivered in 1965.

The lower 3000 series used a 24-bit word size. They were based on the earlier CDC 924 - a 24-bit version of the CDC 1604. The first lower 3000 to be released was the CDC 3200 (May 1964), followed by the smaller CDC 3100 (February 1965), and the CDC 3300 (December 1965). The final machine in the series, the CDC 3500, was released in March 1967 and used integrated circuits instead of discrete components.

All 3000 series computers used magnetic-core memory.

Architecture

The lower 3000 CPU was a 24-bit architecture: instructions were 24 bits in length, as were the two operand registers A and Q. There were four index registers of 15 bits, B0 through B3, though B0 is always zero (zero when read; writes do not affect the value). There was no status (flags or condition code) register. Up to 32,768 words of core memory, 24 bits per word, could be directly addressed, and multiple banks could be switched in. Two or three memory bank configurations were the most common.

Each instruction contained six bits of opcode, one bit specifying whether indirect addressing used, two bits of index register address and fifteen bits of address.

Arithmetic was ones' complement, so there were two forms of zero: positive zero and negative zero. The A and Q register could function as a combined 48-bit register for certain arithmetic instructions. The E register had 48 bits.

The 3300 CPU could execute around one million instructions per second (1 MIPS), giving it supercomputer status in 1965.

Much of the basic architecture design of the 3000 series was done by Seymour Cray, then passed on to others to complete as he moved on to designing the CDC 6000 series. Several of the innovative features that made the 6600 'the first supercomputer' can be seen in prototype in the 3000 series.

Software

The earliest operating system for the lower 3000 series was called RTS OS. FORTRAN (Level 4), COBOL, and ALGOL were available. The assembly language was called COMPASS. These were available from CDC.

Standard peripherals

This is slightly off subject but doesn't appear to exist and is certainly related to the 3000 (and 6000) series of CDC mainframes.

405 - Card Reader. 80 column 'high-speed' punched card reader

415 - Card Punch, 80-column card punch

501 - Line Printer, rotating drum, 136 character wide printer. Note that there were no lower case letters.

505 - Line Printer

512 - Line Printer, chain type

601 - Magnetic Tape Drive

604 - Magnetic Tape Drive

607 - Magnetic Tape Drive

The FORTRAN Language

The bulk of the work was done in FORTRAN (see discussion on Compass later). For anyone well versed in mathematics, FORTRAN was a relatively simple language in which to code.

Again, from Wikipedia:

"FORTRAN, derived from "Formula Translation," is a general-purpose, imperative programming language that is especially suited to numeric computation and scientific computing. Originally developed by IBM in the 1950s for scientific and engineering applications, FORTRAN came to dominate this area of programming early on and has been in continuous use for over half a century in computationally intensive areas such as numerical weather prediction, finite element analysis, computational fluid dynamics, computational physics, crystallography and computational chemistry. It is a popular language for high-performance computing and is used for programs that benchmark and rank the world's fastest supercomputers.

FORTRAN encompasses a lineage of versions, each of which evolved to add extensions to the language while usually retaining compatibility with prior versions. Successive versions have added support for structured programming and processing of character-based data (FORTRAN 77), array programming, modular programming and generic programming (FORTRAN 90), high performance FORTRAN (FORTRAN 95), object-oriented programming (FORTRAN 2003) and concurrent programming (FORTRAN 2008).

In late 1953, John W. Backus submitted a proposal to his superiors at IBM to develop a more practical alternative to assembly language for programming their IBM 704 mainframe computer. Backus' historic FORTRAN team consisted of programmers Richard Goldberg, Sheldon F. Best, Harlan Herrick, Peter Sheridan, Roy Nutt, Robert Nelson, Irving Ziller, Lois Haibt, and David Sayre. Its concepts included easier entry of equations into a computer, an idea developed by J. Halcombe Laning and demonstrated in the Laning and Zierler system of 1952.

A draft specification for The IBM Mathematical Formula Translating System was completed by mid-1954. The first manual for FORTRAN appeared in October 1956, with the first FORTRAN compiler delivered in April 1957. This was the first optimizing compiler, because customers were reluctant to use a high-level programming language unless its compiler could generate code with performance comparable to that of hand-coded assembly language.

While the community was skeptical that this new method could possibly outperform hand coding, it reduced the number of programming statements necessary to operate a machine by a factor of 20, and quickly gained acceptance. John Backus said during a 1979 interview with Think, the IBM employee magazine, 'Much of my work has come from being lazy. I didn't like writing programs, and so, when I was working on the IBM 701, writing programs for computing missile trajectories, I started work on a programming system to make it easier to write programs.'

The language was widely adopted by scientists for writing numerically intensive programs, which encouraged compiler writers to produce compilers that could generate faster and more efficient code. The inclusion of a complex number data type in the language made FORTRAN especially suited to technical applications such as electrical engineering.

By 1960, versions of FORTRAN were available for the IBM 709, 650, 1620, and 7090 computers. Significantly, the increasing popularity of FORTRAN spurred competing computer manufacturers to provide FORTRAN compilers for their machines, so that by 1963 over 40 FORTRAN compilers existed. For these reasons, FORTRAN is considered to be the first widely used programming language supported across a variety of computer architectures.

The development of FORTRAN paralleled the early evolution of compiler technology, and many advances in the theory and design of compilers were specifically motivated by the need to generate efficient code for FORTRAN programs."

A simplified flow of the "debugging process"

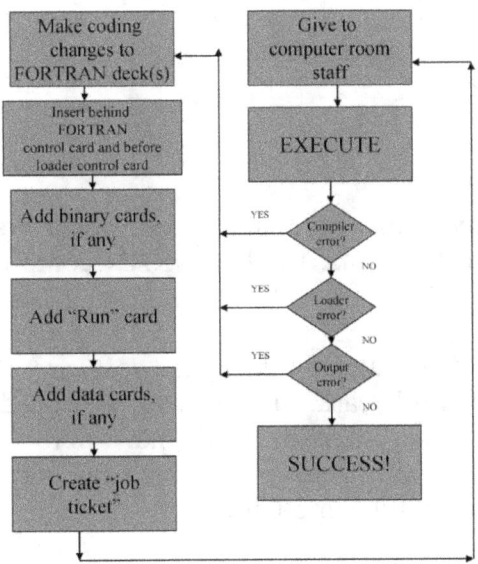

The first step was to create or edit the FORTRAN logic being tested. If the number of cards to be keypunched was "large," the FORTRAN logic was hand-written on a coding sheet (see below) and given to a 026-keypunch operator.

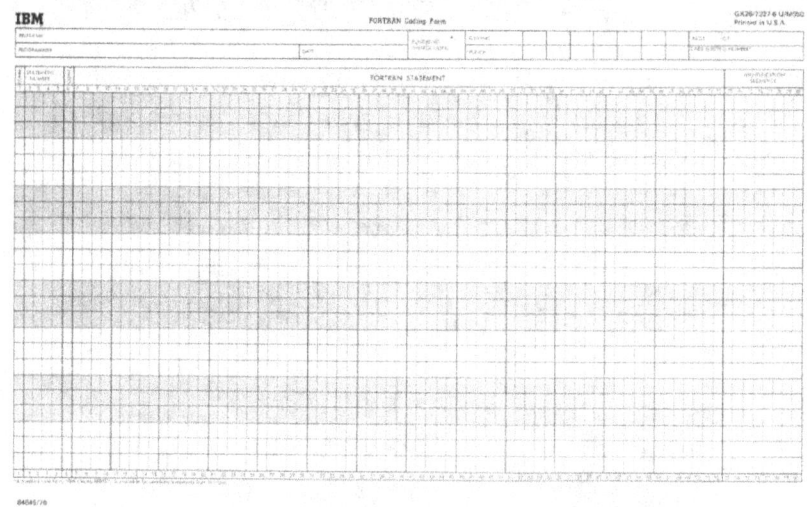

Step 2: The FORTRAN source code, now ready to test, was inserted between a control card to invoke the FORTRAN compiler and the control card to invoke to loader.

Step 3: Any binary cards for already-debugged software were placed behind the loader control card. There was no need to have FORTRAN recompile theses routines if they were "bug-free." FORTRAN produces a "binary deck" of machine language code whenever there is a successful compilation.

Step 4: The "Run" card was inserted, indicating the end of the binary decks.

Step 5: If there were data cards required as input for the FORTRAN programs, they would go behind the "Run" card.

Step 6: A "job ticket" was created to go with the card deck to the computer room. This showed the programmer's ID, the job's priority, and any special instructions.

Step 7: The job went to the computer room.

Step 8: The job was run.

Step 9: If there were any FORTRAN compilation errors, they would be indicated on the program listing on the line printer. If so, the job would abort here.

Step 10: If there were any loader errors, they would be indicated on the storage map on the line printer. Typically, any "undefined references" constitute loader errors (a missing subroutine, a misspelled name, etc.).

Step 11: The printer, magnetic tape, and plotter outputs (if any) were carefully examined. Success, if no errors in all the outputs ... but this would take months.

A simple FORTRAN program example (Wikipedia)

Multiple data card input

This program has two input checks: one for a blank card to indicate end-of-data, and the other for a zero value within the input data. Either condition causes a message to be printed.

```
C AREA OF A TRIANGLE - HERON'S FORMULA
C INPUT - CARD READER UNIT 5, INTEGER INPUT, ONE BLANK CARD FOR END-OF-DATA
C OUTPUT - LINE PRINTER UNIT 6, REAL OUTPUT
C INPUT ERROR DISPAY ERROR MESSAGE ON OUTPUT
  501 FORMAT(3I5)
  601 FORMAT(4H A= ,I5,5H  B= ,I5,5H  C= ,I5,8H  AREA= ,F10.2,
     $13H SQUARE UNITS)
  602 FORMAT(10HNORMAL END)
  603 FORMAT(23HINPUT ERROR, ZERO VALUE)
     INTEGER A,B,C
  10 READ(5,501) A,B,C
     IF(A.EQ.0 .AND. B.EQ.0 .AND. C.EQ.0) GO TO 50
     IF(A.EQ.0 .OR.  B.EQ.0 .OR.  C.EQ.0) GO TO 90
```

```
      S = (A + B + C) / 2.0
      AREA = SQRT( S * (S - A) * (S - B) * (S - C) )
      WRITE(6,601) A,B,C,AREA
      GO TO 10
   50 WRITE(6,602)
      STOP
   90 WRITE(6,603)
      STOP
      END
```

Let's look at each instruction, one by one. The cards starting with the letter "C" are comments. After this are four format statements that indicate how the input or output is to be formatted. "3I5" means three consecutive five-digit integers. "xH" indicates x characters follow the "H" (for printing purposes). "I5" again, means a five-digit integer and "F10.2" means a computed ("floating point") number within a ten-character field with two decimal places of precision.

The first "executable" statement is statement 10; it says, "read one card, three integers (A, B, and C) in five-digit format.

The next statement says, "if A and B and C are zero (blank)," go to statement 60, which prints out "NORMAL END."

The next statement says, "if A OR B OR C equals zero, go to statement 90 which prints out "INPUT ERROR, ZERO VALUE."

If the inputs were valid, the computation takes place, computing the area of the triangle and saving it as "AREA." The answer is then printed using format 601, and the program branches to statement 10 to read yet another card.

The FORTRAN language is, of course, much richer than in the illustrated example but, it is hoped, the example provides "a taste."

Down in the dumps

Not all programs end gracefully with an error message or successful outputs. Sometimes a program will go into an endless "loop" or come to a crashing halt. When this happens, the resourceful computer operator will go to the CDC 3200's console and write down all the computer's registers and the current memory address. The computer operator then does a "core dump", printing out the entire contents (32,768 words) of memory on the line printer (taking just a few minutes). The computer operator's hand-written comments, along with the core dump printout, the FORTRAN compiler listing, the loader map, and any outputs go back to the programmer who submitted the job. Analyzing a core dump was enjoyable for me; it was like solving a puzzle.

Symbolic map of CDC 3200 memory

CDC "operating system"
Main FORTRAN program
FORTRAN subroutines
FORTRAN runtime library entries
Workspace (if any)
COMMON area

The CDC operating system, RT OS, occupied lower memory. This contained the device drivers, etc. The main FORTRAN program followed next, followed by any FORTRAN subroutines. Finally, the CDC loader satisfied any external references by pulling in entries from the FORTRAN library – routines such as SINE, COSINE, etc.

The rear portion of memory was FORTRAN's COMMON area. In order to save space and maximize the use of memory, programmers would place all of their key variables in the COMMON area. These variables were then shared by all the programmer's subroutines. It was important for each FORTRAN program and subroutine to have exactly the same COMMON area allocation because if they were not aligned perfectly, one program would potentially overwrite the variables of another program – a common programming error.

Between the end of the program space and the COMMON area, there was workspace. Because the computer memory was so small, it frequently happened, as programs got large, that there was no workspace and the program space overlaid the COMMON area. When this happened, the programmer needed to reduce the size of the program or beak it up into "overlays" that would come off a magnetic tape and flow into the program space, leaving the COMMON area untouched.

In designing the software architecture for MP^2, I made a decision to have the mission's data reside in COMMON, while the programs shuttled in and out from an overlay tape. None of this, of course, would be necessary if disk storage had been available!

Even FORTRAN had its flaws

We would have hoped that FORTRAN IV was totally "bug-free," but such was not the case.

There were instances (I do not recall the specifics) when certain FORTRAN run-time routines (SINE, e.g.) needed to be modified because they caused our software to crash if they had an erroneous input. To remedy this situation, I sometimes had to request the source code for a specific run-time library program from Control Data and reassemble it, with my modifications, to handle these conditions. This did not happen frequently, but it was a major disruption when it did happen. To my recollection, these instances where the only times I had to use CDC's Compass assembler program.

And now, MP^2...

Our mission-planning software (proprietary, of course) had three distinct segments:

1. Flight Path: Given a mission start point and mission end point, simulate the flight of the SR-71 while incorporating the necessary refuelings.

2. Ground track: Given a set of potential "points of interest" and a set of "avoidance areas," create an optimum swath for the sensors to incorporate as many points of interest as possible while avoiding the "avoidance areas." This was a very sophisticated program, often referred to as the solution to the "traveling salesman problem." With a complex mission and many points of interest, the software could literally run for hours.

3. Output Phase: The flight path, points of interest, and avoidance areas were all plotted on a flatbed plotter. A Mylar tape was produced for the Nortronics Navigation System.

What type of mathematics was required?

Aircraft navigation heavily utilized "great circles." Referring again to Wikipedia:

> A great circle track is the shortest distance between two points on the surface of a sphere; the Earth isn't exactly spherical, but the formulas for a sphere are simpler and are often accurate enough for navigation.

> Course and distance [edit]
>
> The great circle path may be found using spherical trigonometry; this is the spherical version of the inverse geodesic problem. If a navigator begins at $P_1 = (\phi_1, \lambda_1)$ and plans to travel the great circle to a point at point $P_2 = (\phi_2, \lambda_2)$ (see Fig. 1, ϕ is the latitude, positive northward, and λ is the longitude, positive eastward), the initial and final courses α_1 and α_2 are given by formulas for solving a spherical triangle:
>
> $$\tan \alpha_1 = \frac{\sin \lambda_{12}}{\cos \phi_1 \tan \phi_2 - \sin \phi_1 \cos \lambda_{12}}$$
>
> $$\tan \alpha_2 = \frac{\sin \lambda_{12}}{-\cos \phi_2 \tan \phi_1 + \sin \phi_2 \cos \lambda_{12}}$$
>
> where $\lambda_{12} = \lambda_2 - \lambda_1$ and the quadrants of α_1, α_2 are determined by the signs of the numerator and denominator in the tangent formulas (e.g. using the atan2 function). The central angle between the two points, σ_{12}, is given by
>
> $$\cos \sigma_{12} = \sin \phi_1 \sin \phi_2 + \cos \phi_1 \cos \phi_2 \cos \lambda_{12}.$$
>
> The distance along the great circle will then be $s_{12} = R\sigma_{12}$, where R is the assumed radius of the earth and σ_{12} is expressed in radians. Using the mean earth radius, $R = R_1$ yields results for the distance s_{12} which are within 1% of the geodesic distance for the WGS84 ellipsoid.

Finding way-points

To find the way-points, that is the positions of selected points on the great circle between P_1 and P_2, we first extrapolate the great circle back to its node A, the point at which the great circle crosses the equator in the northward direction; let the longitude of this point be λ_0 — see Fig 1. The azimuth at this point, α_0, is given by the spherical sine rule:

$$\sin \alpha_0 = \sin \alpha_1 \cos \phi_1.$$

Let the angular distances along the great circle from A to P_1 and P_2 be σ_{01} and σ_{02} respectively. Then using Napier's rules we have

$$\tan \sigma_{01} = \frac{\tan \phi_1}{\cos \alpha_1}. \quad \text{(If } \phi_1 = 0 \text{ and } \alpha_1 = \tfrac{1}{2}\pi, \text{ use } \sigma_{01} = 0.)$$

This gives σ_{01}, whence $\sigma_{02} = \sigma_{01} + \sigma_{12}$.

The longitude at the node is found from

$$\tan \lambda_{01} = \frac{\sin \alpha_0 \sin \sigma_{01}}{\cos \sigma_{01}},$$
$$\lambda_0 = \lambda_1 - \lambda_{01}.$$

Finally, calculate the position and azimuth at an arbitrary point, P (see Fig. 2), by the spherical version of the direct geodesic problem. Napier's rules give

$$\sin \phi = \cos \alpha_0 \sin \sigma,$$
$$\tan(\lambda - \lambda_0) = \frac{\sin \alpha_0 \sin \sigma}{\cos \sigma},$$
$$\tan \alpha = \frac{\tan \alpha_0}{\cos \sigma}.$$

The atan2 function should be used to determine σ_{01}, λ, and α. For example, to find the midpoint of the path, substitute $\sigma = \tfrac{1}{2}(\sigma_{01} + \sigma_{02})$; alternatively to find the point a distance d from the starting point, take $\sigma = \sigma_{01} + d/R$. Likewise, the vertex, the point on the great circle with greatest latitude, is found by substituting $\sigma = +\tfrac{1}{2}\pi$. It may be convenient to parameterize the route in terms of the longitude using

$$\tan \phi = \cot \alpha_0 \sin(\lambda - \lambda_0).$$

Latitudes at regular intervals of longitude can be found and the resulting positions transferred to the Mercator chart allowing the great circle to be approximated by a series of rhumb lines. The path determined in this way gives the great ellipse joining the end points, provided the coordinates (ϕ, λ) are interpreted as geographic coordinates on the ellipsoid.

These formulas apply to a spherical model of the earth. They are also used in solving for the great circle on the auxiliary sphere which is a device for finding the shortest path, or geodesic, on an ellipsoid of revolution; see the article on geodesics on an ellipsoid.

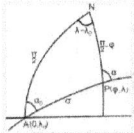

Figure 2: The great circle path between a node (an equator crossing) and an arbitrary point (ϕ, λ).

Example

Compute the great circle route from Valparaíso, $\phi_1 = -33°$, $\lambda_1 = -71.6°$, to Shanghai, $\phi_2 = 31.4°$, $\lambda_2 = 121.8°$.

The formulas for course and distance give $\lambda_{12} = -168.6°$, [note 7] $\alpha_1 = -94.41°$, $\alpha_2 = -78.42°$, and $\sigma_{12} = 168.55°$. Taking the earth radius to be $R = 6371$ km, the distance is $s_{12} = 18743$ km.

To compute points along the route, first find $\alpha_0 = -56.74°$, $\sigma_1 = -96.76°$, $\sigma_2 = 71.8°$, $\lambda_{01} = 98.07°$, and $\lambda_0 = -169.67°$. Then to compute the midpoint of the route (for example), take $\sigma = \tfrac{1}{2}(\sigma_1 + \sigma_2) = -12.48°$, and solve for $\phi = -6.81°$, $\lambda = -159.18°$, and $\alpha = -57.36°$.

If the geodesic is computed accurately on the WGS84 ellipsoid, the results are $\alpha_1 = -94.82°$, $\alpha_2 = -78.29°$, and $s_{12} = 18752$ km. The midpoint of the geodesic is $\phi = -7.07°$, $\lambda = -159.31°$, $\alpha = -57.45°$.

Alpha testing at Edwards Air Force Base

In 1965 and after my security clearance, I received word from Marv Goldstein that I'd be "going to L.A." Being naïve and never having flown on a commercial flight, I rushed home to tell my wife that we'd soon be going to Louisiana (LA). How wrong I was!

In that year, the SR-71 was being tested at Edwards Air Force Base before moving it to Beale Air Force Base in January of 1966. Also, in 1965 a subset of the MP² software was being tested at Edwards as a show of capabilities. As I remember it, the MP² subset was basically what was later called "Flight Path," together with the plotter output. All of this, the CDC 3200 and its peripheral devices, together with the flatbed plotter, were housed in a bright blue air-conditioned trailer out in the middle of a runway.

I know I made a few trips to Edwards in 1965. I stayed in Lancaster, California and commuted to Edwards AFB out in the Mojave Desert, a distance of thirty-three miles. It was blazing hot. I remember staying in a Lancaster motel room and being awakened by the tremors of a "light" earthquake. I remember driving through the desert and looking at the weird Joshua trees.

For those of you who have never heard of Edwards Air Force base, here is the Wikipedia overview:

> "It is the home of the Air Force Test Center and is the Air Force Materiel Command center of excellence for conducting and supporting research and development of flight, as well as testing and evaluation of aerospace systems from concept to combat. It operates the U.S. Air Force Test Pilot School and is home to NASA's Armstrong Flight Research Center and considerable test activities conducted by America's commercial aerospace industry.

Previously known as Muroc Air Force Base, Edwards AFB is named in honor of Captain Glen Edwards (1918–1948). During World War II, he flew 50 missions in A-20 Havoc light attack bombers in the North African campaign on extremely hazardous, low-level missions against German tanks, convoys, troop concentrations, bridges, airfields, and a variety of other tactical targets. Edwards became a test pilot in 1943 and spent much of his time at Muroc Army Air Field, on California's high desert, testing wide varieties of experimental prototype aircraft. He died in the crash of a Northrop YB-49 flying wing near Muroc AFB on 5 June 1948.

The base is next to Rogers Dry Lake, an endorheic desert saltpan whose hard dry lake surface provides a natural extension to Edwards' runways. This large landing area, combined with excellent year-round weather, makes the base good for flight-testing. The lake is a National Historic Landmark.

The base has played a significant role in the development of virtually every aircraft to enter the Air Force inventory since World War II. Almost every United States military aircraft since the 1950s has been at least partially tested at Edwards, and it has been the site of many aviation breakthroughs.

Notable occurrences at Edwards include Chuck Yeager's flight that broke the sound barrier in the Bell X-1, test flights of the North American X-15, the first landings of the Space Shuttle, and the 1986 around-the-world flight of the Rutan Voyager."

The trailer in which we worked was cramped; it was also difficult to keep it air-conditioned in one hundred-degree plus temperatures. If we needed to go to the bathroom, we had to cross the runway and enter the SR-71 hangar. I was able, in those instances, to give the Blackbird a loving pat on the right engine.

I think I was good at "debugging." I know that I was testing not just the software I had written, but also the software of a number of other programmers. I remember taking quite a few core dumps back to the motel for analysis.

On one of the trips, I went out to Lancaster with my wife, Amy, and son, Adam. We were able to drive around in the desert as well as go into Los Angeles to visit the La Brea Tar Pits, Warner Brothers Studio, Hollywood, Malibu, and some excellent Jewish delis.

On another trip, I was able to drive into the Mojave Desert and fly a glider. If one looks up at the sky to view a glider in motion, it looks smooth and peaceful. Such is NOT the case. The tow-plane took us (I was with a pilot) up to altitude and cut us loose. We rode the thermals and coasted back down. It was a very, very bumpy ride which, together with the intense heat, brought me to the point of nausea.

All ended well later that year. The Air Force accepted our alpha-phase software and we returned to Deer Park, New York to start the full-blown MP^2 software.

Beta Testing at Beale AFB

If 1965 was the year for alpha testing at Edwards, 1966 was the year for integration of the full-blown MP^2 software and the beta test for the Air Force. The first SR-71 had arrived at Beale in January of that year.

> ***From Wikipedia:*** *"Beale AFB is the home of the 9th Reconnaissance Wing (9 RW), which also serves as the host wing for the base. The installation is located outside of Linda, about 10 miles (16 km) east of the towns of Marysville and Yuba City and about 40 miles (64 km) north of Sacramento. Beale is a large base in terms of land and has five gates providing access on all sides of the base."*

I made many trips to Beale in 1966. In the spring of that year, Joe Cooper and Joel Raphael (see their chapters) went out to Sacramento to secure apartments for nine of the team's families, which included their wives and children. Although we all worked hard during the week, that summer provided me with one of the most memorable experiences of my life. We travelled to San Francisco, Carmel, Monterrey, Yosemite, Reno, Lake Tahoe, Napa Valley, and even to Tijuana.

There was one time when I started running a very high fever and had a splitting headache. My wife took me to the ER, where they did a spinal tap, looking for signs of meningitis. The spinal tap was excruciating, but there was no meningitis – in fact they never figured out what it was. The headache persisted for several weeks. During the first week or two, I just had to lie down and rest by the pool, often examining core dumps. Marv Goldstein visited the hospital while I was there and brought me a large (gallon?) bottle of Jim

Beam – not the best cure for a headache, but ultimately very enjoyable.

We worked in the SAGE Building at Beale. I do not remember the building all that well, but I do recall always having to be escorted to the men's room (for security purposes).

The evenings were enjoyable. We'd arrive late and sit by the pool. Of course, there was no "shop talk" about work; we maintained whatever cover stories we could conjure up. I recall one weekend at poolside in which my two-year-old son, Adam, was racing along the edge of the pool, ultimately falling in. I dove in after him … it was a while before we got him interested in learning how to swim.

The fall came and many of the families went home, including mine. I still made a number of trips to Beale afterwards and, ultimately, the Air Force became involved and we started the demonstrations of MP^2 and the steps toward acceptance testing.

I don't remember if this was actually a demonstration with the Air Force present or just a "dress rehearsal." In any event, one of the very complex test cases, with many "points of interest," ran for hours on the 3200, ultimately causing it to crash with a memory parity error. A memory parity error is a computer hardware error that should never occur with any level of memory thrashing. It did occur, and it occurred several times, never in the same place. I don't know if we gave up on that mission or just changed the data stream.

The IBM 7090 at SAC Headquarters

I believe this was in 1967. Part of the Air Force contract was a commitment to get the CDC 3200 FORTRAN IV code converted over to the IBM 7090 at SAC Headquarters in Omaha,

I made a few trips to Omaha and they, also, were memorable. Constantly escorted, I'd get in an elevator and go down many levels until we reached the glassed-in room with the IBM 7090. We struggled mightily with the conversion. You'd think that FORTRAN equals FORTRAN, but such was not the case; the two FORTRANs were not at all compatible.

The direct conversion did not work, so we wrote a "filter program" to convert the one FORTRAN into the other. That also had its difficulties.

I left Airborne before this effort concluded. To this day, and after many inquiries, I still do not know if MP^2 ever ran on the 7090.

Recollections of five of the Airborne Instruments project team

I was able to locate (not a trivial task) five of the team with whom I worked fifty-plus years ago. I interviewed each of them about their experiences. All of them had less than perfect recall, but here are their stories:

An Interview with George Adjami – From Bell Labs to the SR-71, then the LEM, and later real estate

George graduated with a BS degree in mathematics from Brooklyn College, and in 1963 went to work as a systems/programmer analyst at Bell Labs in Whippany New Jersey. Under a Work/Study Program, Bell Labs allowed George to further his education and work towards a master's degree at Stevens Institute of Technology. The program allowed him to work on-the-job three days a week and attend Stevens Institute the other two days. He later completed and obtained his master's degree in Computer Science from Adelphi University.

At Bell Labs, George worked on the Nike-X/Nike-Zeus ABM (anti-ballistic missile) system. It was designed for the Army to destroy ICBM warheads before they could hit large city targets in the U.S. The later Nike-X system was developed to overcome the limitations of the earlier Nike-Zeus, and to provide the needed speed and accuracy, as well as deal with multi-warhead attacks. Nike-X used a new radar system and high-speed computers that could track hundreds of objects at once. George utilized his math, FORTRAN, and assembly language skills to develop data reduction and analysis programs to analyze simulated data.

In 1965, with the aerospace industry booming, George, along with his wife and newborn son, decided to relocate back to Long Island, New York and begin a new job opportunity at Airborne Instruments Laboratory. He interviewed with John Boccio, then the Programming Manager, and with Marvin Goldstein, who functioned as an assistant project manager. In George's words, "This was my second job out of college and the state-of-the-art SR-71 program and the AIL staff working on the project were of great interest to me and were a great fit for my technical background and computer skills."

George wrote FORTRAN programs at both the main program and the subroutine level. He worked closely with a team of systems analysts and engineers who formulated the logic that would be the basis for the programming code. The programs were written for The CDC 3200 FORTRAN compiler that took magnetic tape and "punched cards" as input.

George was working on what we called "Flight Path" – the software that computed the SR-71 flight plan. He never went out to Edwards Air Force Base for the "alpha test" but, in the summer of 1966, joined the large team doing the final integration for the "beta test" at Beale Air Force Base. Initially AIL wanted to send out just the programming and systems staff, but after a few discussions, management agreed that their spouses and children could join the staff for a summer in Sacramento, California. Of course, no one could tell his or her families the precise nature of the work being done or the installation location. Still, this was acceptable to the families and everyone had an amazing summer in California, living in nicely furnished apartments and socializing with the other families.

George still talks about the summer of 1966 to this day. "We all worked hard, but the weekends were special." The team gathered every Friday night to plan weekend activities. The team got to see places like Reno, Lake Tahoe, San Francisco, Yosemite, Muir National Park, Monterrey, Carmel, Napa Valley, and even Tijuana. At the time, George was there with his wife Evelina and two-year-old son, Mark. "The years at Airborne were fulfilling, rewarding, and fun. What I liked most about the project was the team spirit."

An ambitious person by nature, George wanted to take advantage of the hot aerospace job market and so, in 1967, he left AIL and went to work at Grumman Aerospace as a software consultant working on the LEM (Lunar Excursion Module). "The project was in its full development phase, and the thought of working on a spacecraft that went to the moon had a great deal of visibility, significance, and technical challenge," George said.

Grumman was run like a family company and the employees referred to themselves as "Grummanites." They worked hard and long hours and desperately wanted to meet president Kennedy's deadline of putting a man on the moon before the 1960s ended. George recalls, "It was stressful at times working a 50-60-hour week, including weekends, to meet project deadlines and mission objectives. But no one seemed to mind it and everyone enjoyed the overtime pay." George worked closely with his counterpart engineers to develop and test software for the various instruments and sub-systems of the Lunar Module.

George sums it up as follows: "It was a very exciting time. Everyone worked as a team and was motivated to pitch in and support each other to make the project a success." He remembers the day in July 1969 when the "Eagle" landed on the moon. "We were in Bethpage watching television and we were amazed at what was going on, but also worried that something might go wrong. When the LM landed, there was a great feeling of joy, satisfaction, and a sense of significant accomplishment for Grumman, NASA, and the country. Even today, there is a group of retirees that meet and share stories of the LM program and the memorable July 1969 lunar landing."

There was a brief period (1969-1971) in which George formed a software consulting company (Digiscan) with John Altson and Roger Mazur. The company provided programming and systems analysis expertise, as well as personnel to work on numerous projects for clients in the scientific, aerospace and commercial marketplace. In 1971, the company experienced the downtrend in the economy and the demand for its services declined. That, along with the financial backers being at odds with the three principals, forced the company to slowly wind down and eventually close the business.

In 1971, George went back to Grumman as a full-time employee, working for the vice-president of the newly formed Grumman Data Systems Division. With the advent of mini-computers, George headed up a mini/micro computer group. They worked with many diverse state-of-the-art mini computer platforms including the Perkin Elmer/Interdata, Digital Equipment, Data General, DataPoint, and other computer manufacturer platforms. The mini/micro group provided their expertise and services to numerous internal Grumman Aerospace proposals and projects, as well as numerous projects for commercial clients.

In 1984, George left the aerospace industry and moved on to the business/commercial world. He joined Data General (a mini computer manufacturer) and managed the systems engineering group for both the Wall Street branch and the Midtown branch office in New York City. He worked closely with the sales team in meeting both revenue and client penetration objectives. George said, "It was a very competitive/hectic environment involving both pre-sales and post-sales activities. There was constant pressure to meet revenue targets and delivery schedules." Clients included numerous Wall Street brokerage/financial firms, insurance firms, and the Federal Reserve. In early 1990, Data General merged with EMC Corp and left the mini computer market to specialize in mass disk data storage systems instead.

In 1991, it was time for George to exit the commercial rat race and work in a "user-environment." He joined the Medical Society of the State of New York (MSSNY) as Vice President of Technology and Physician Services. He worked out of the MSSNY headquarter office in Lake Success, New York for over 17 years. George stated, "The Medical Society provided all the services of a non-profit Association including advocacy, membership development and retention, marketing, continued medical education, and member benefits. MSSNY was part of a national network of separately managed state, county, and local medical societies." George interacted with many of these state/local societies, as well as the AMA at the national level.

His Division provided the entire computer, network, and technology-related needs and services to the MSSNY staff in Lake Success, as well as the Albany office. His group maintained the MSSNY website and an up-to-date online physician database that can be accessed by member physicians, hospitals, other state and local medical societies, as well as the public. The database included biographical, medical training, specialty, and other pertinent data for over 70,000 New York State physicians, as well as medical students. His Division was also responsible for publishing the bi-annual *Medical Directory of the State of New York* (a reference publication that is used by physicians and available to the public at most public libraries).

In 2007, and at the age of 66, George "retired." He had, by then, remarried. Along with Roberta ("Bobbi'), his current wife of 30 years, he was about to start a life of leisure.

His son, however, had a successful real estate business and asked George to come onboard. The two of them were successful together, taking foreclosure properties from banks - managing and fixing the properties on behalf of the banks and preparing them for resale. In many instances, they had ready cash buyers, investors and developers for the resale. In addition, both he and his son have invested in a number of properties themselves, to fix up, resell and flip or hold on to as rental income.

When George's wife, Bobbi, retired from the teaching profession in 2011, George slowed down his real estate work and focused on his family – a son, a daughter, and five grandchildren. His hobbies include golf, fishing and investing in the stock market. George and his wife like to travel and have gone on numerous cruises. "Life is good, and we are enjoying every bit of it," said George at the conclusion of the interview.

.

Mel Berger – From the SLR Logic to AT&T Labs

Airborne Instruments Labs hired Mel in 1964. He was briefly on a few short projects and then moved to a secure area for an undisclosed project – SR-71 mission-planning.

After he received his Top-Secret clearance, Mel got involved in learning the Side-Looking-Radar (SLR); as a systems analyst he developed the logic for that sensor for use onboard the SR-71, developing flowcharts and requirements for the programming team. The work was "very interesting" and "a fun project on which to work."

The SLR was used in the "Ground Track" phase, and, for his test cases, he was never told the destination of the missions.

Mel's California travels

Although he never went to Edwards Air Force Base, he was one of nine families spending the summer of 1966 in Sacramento; he commuted each weekday from Sacramento to Beale AFB. Mel even went out with Joe Cooper early in the summer to scout out the housing (and supplies) for the nine families.

Mel's first wife was with him that summer, and they enjoyed their weekend jaunts to San Francisco, Palo Alto, Monterey, Yosemite, Muir Woods and other surrounding areas. Mel was asked to stay beyond the summer, but he declined – having to return to New York to finish his master's degree.

Mel enjoying California

Two stories

When Mel arrived at the Beale gatehouse, he had to be escorted to the SAGE Building. It was a long walk and the escort was moving very slowly. Mel and his teammates had tight deadlines, so Mel asked him why the slow pace. The escort said that he had four years left at the base and he was in no hurry.

On another occasion, Mel and Joel Raphael were stuck in the SAGE building's elevator. Fortunately, there was a phone in the elevator, but it seemed like an eternity before they were "rescued." (The AIL team members always needed to be escorted to the bathrooms in the SAGE Building).

Reflections, and moving on...

Mel commented that the cell phones of today are more powerful than the CDC 3200 used for software development – a tribute to technological advances over the fifty elapsed years!

In January of 1967, the project was winding down. Mel was "not learning anything new" so it was time to move on. He went to work at Sperry Rand on Long Island doing "very interesting systems engineering work" on "inertial navigation" (stabilizing the movements through the ocean) to improve the targeting accuracy of submarines that launched the Polaris and Poseidon Missiles. He "got to use the advanced mathematics" that he had learned in graduate school.

Mel found that the Defense Industry was shaky, so he looked for a change. He picked up his master's degree and, after constant enticements from his friend Joel Raphael, went to work at Bell Labs, the R&D Division of AT&T.

At Bell Labs, Mel worked first as a systems engineer, and then as a systems engineering manager on various communications services projects (e.g., Calling Card, Advanced 800, Voice Messaging, and Voice Calling services that used advanced speech recognition technology). He continued that work at AT&T Labs after the spin-off of Lucent Technologies, which retained Bell Labs. Mel worked at AT&T for thirty-five years, retiring as a Division Manager in 2003.

Mel and his second wife now live in New Jersey and in Florida, sharing time in two different retirement communities. There are always "fun things to do" (e.g., playing Pickle Ball and golf, and enjoying Ballroom dancing).

He has two daughters from his first marriage and a stepdaughter and stepson. He and his wife have nine grandchildren (five girls and four boys, ages five through twenty) with whom they enjoy spending time.

Joe Cooper – A man wearing many hats

Joe graduated from college with a degree in aeronautical engineering and was hired by the Glenn L. Martin Company in Baltimore, Maryland in 1956. He learned how to program the IBM 1620 computer and was assigned to develop the pilot's flight handbook for the B57-D, the first spy plane before the U-2. As part of that task, he used his knowledge of the 1620 computer to determine the rate of climb data for use in the handbook. The project required his obtaining a Top-Secret clearance.

When the Martin Company left the airplane business and went into electronics, Joe, acting on a friend's advice, attended Johns Hopkins University where he was awarded a bachelor's degree in electrical engineering. Then, in 1964, Martin merged with TRW to become the Bunker Ramo Corp. Joe was caught up in the change and continued doing cryptographic analyses of NSA boxes for the new company, making sure they could not be hacked. This work required Joe's Top-Secret clearance. His group moved to Silver Spring, Maryland, and after a while Joe decided to leave Bunker Ramo.

Airborne Instruments Laboratory hired Joe because of his education, background, and Top-Secret clearance. He was slated to work on ELINT sensors. He knew absolutely nothing about the SR-71, but that quickly changed when he was transferred to the SR-71 project, nicknamed MP^2 (Mission-Planning squared, where the second "P" stood for (mission) Preparation for the output phase – the creation of the ANS Mylar tape to fly the SR-71 and the filmstrip for the RSO to watch the mission).

The "very crude" solution

Joe was tasked with figuring out how to create the filmstrip for the RSO. First, he bought the top of the line Exakta camera. He pulled out the viewfinder and inserted a light and reticle in its place. He brought the modified camera into the model shop and had the shutter filed open from the 35M area to create continuous film.

Then Joe went to the ACIC (Aeronautical Chart and Information Center) in St. Louis to learn all about maps. He purchased the maps of the United States and brought them back to AIL where, with the use of a tripod, he carefully photographed a simulated mission and created the first test mission filmstrip. He traveled to SAC Headquarters in Omaha, Nebraska to show the product and was given a nod of approval by the Air Force.

Joe's next step was a visit Disney to assess the feasibility of using animation cameras to prepare mission filmstrips. He also researched flatbed plotters and, finally, went back to ACIC for more maps and equations.

The call to Edwards Air Force Base (1965)

The manager for AIL's SR-71 project was Mike Ettengoff. Mike asked Joe to drop what he was doing and go to Edwards to assess the status of the "alpha testing" of the MP^2 software. Joe flew out to Edwards and found that schedules were not being met. When he returned and gave Mike the status, Mike put Joe in charge of the Edwards implementation effort to make sure the integration would go well.

One major aspect of the MP^2 software was the solution to "The Traveling Salesman Problem." A very talented PhD, Dick Springer, and a young programmer, Joyce Palmer, were assigned to this problem that, in essence, was creating an SR-71 route that would maximize the number of targets while circumventing the "avoidance areas" such as enemy radar. Joe worked with Dick to make sure that the SR-71 aircraft dynamics were properly incorporated into the software. (Additionally, Joe had to suggest that Joyce stop using her boyfriends' names as variables in her programs).

There were many hardware problems on the CDC 3200 computer installed in the Mobile Processing Center (MPC); the desert heat may have been a factor. Joe smelled fire one day from an undetermined source. He told the hardware serviceman to "let it burn," which they did, so that they could isolate the malfunctioning electronics within the tape drive. The "let it burn" nickname stayed with Joe throughout the remainder of the development.

Joe made several trips to SAC HQ in Omaha to advise the Air Force brass on the status at Edwards.

And then, the beta testing at Beale Air Force Base (1966)

Joe told Mike Ettengoff that he'd supervise the project at Beale if and only if he was granted control of the (expense) finances. He was given that latitude and traveled with a small party to set up apartments in Sacramento, making sure that they were equipped with television sets, house cleaning, and furniture. All was now ready for the full AIL team to arrive. They commuted 80 miles every day between Sacramento and Beale Air Force Base in Marysville, California.

The team worked for the entire summer of 1966 debugging the already-integrated software. The Disney camera and flatbed plotter were tested in the MPC. Now the team was ready for preparing test flights, one of the first being a mission over Cuba.

Joe's "Top-Secret" coffee cup and SR-71 model

Joe's Farewell to AIL

Merv Weich, one of Joe's co-workers, had taken a job at the Raytheon Company in Massachusetts. In 1967, he invited Joe to a lunch in San Francisco that turned out to be a Raytheon interview for a job working on an en-route air traffic control system bid. Joe did not give an answer then and went back to the SR-71 project. During a bitter disagreement about the way the project's status was being conveyed to the Air Force, Joe resigned from AIL and accepted the Raytheon job in Boston.

Joe's twenty-eight years at Raytheon

Initially, Joe went to work on the air traffic control system. He devised the logic for the display systems and, during that same time, worked on computerizing airline reservation systems. He was also involved with many of Raytheon's large radar systems, such as Aegis, Cobra Dane, NATO Sea Sparrow and the Strategic Defense Initiative ("Star Wars") missile defense system.

It happened that a Raytheon executive was on a flight with some executives from the Washington Post. The Washington Post and a consortium of seven newspapers were keenly interested in Raytheon converting the display system for air traffic control into a newspaper pagination system for their composing room. Just like that, Joe was on a new career track.

Joe spent several days working at the (Norfolk) Virginian-pilot, acting as a newspaper employee and learning the flow of information. The Washington Post group had awarded Raytheon a large contract to develop the pagination system for advertising makeup using the Norfolk, Virginia newspaper for its pilot program. Joe also worked with Washington Post executives to develop a unique and advanced editorial system employing the latest fault-tolerant computers and displays for the newsroom (the same newsroom depicted in the film "All the President's Men").

The project was successful, but at a development cost of eleven million dollars. The contract from the Washington Post was five million dollars and Raytheon frowned upon Joe's "charitable effort" to that newspaper. Unfortunately, Raytheon failed to realize that they now had a product that was in demand at the Los Angeles Times, Detroit Free Press, Miami Herald and other large newspapers; it could have been sold many times over at a large profit. At that time, under Reagan, the defense industry caught fire, so a risk-averse Raytheon sold off its newspaper business and went back to its comfortable cost-plus 6% profit business model in the defense industry.

It was during this newspaper phase at Raytheon that Joe became a much sought-after speaker at newspaper publishers' conferences.

After the newspaper experience, Joe was involved in marketing microwave and advanced technologies, as developed by a small Raytheon R&D team, to Amana, Caloric and other Raytheon commercial companies. Joe said that this endeavor was "not too effective."

Joe's final position at Raytheon was Marketing Manager for the engineering organization, a force of several thousand engineers within the Equipment Division. During the next twelve years, he was responsible for administering research and development projects and technology contracts for the military. In 1995, at age 60, and after 28 years at Raytheon, Joe retired and moved with his wife, Linda, to Hilton Head Island, South Carolina, where his former engineering manager and family had moved several years earlier. They stayed in touch as friends and golfing partners.

In 2017, on a weekend vacation in Manhattan, Joe and Linda visited the Intrepid, docked in New York's Hudson River at 42nd Street, and were able to stand alongside the SR-71 that is displayed on the ship.

Joe's Civil Activities

When Joe accepted the job offer at Raytheon, he and his family moved to Framingham, Massachusetts, where he soon involved himself in auxiliary activities.

For 17 years, Joe was an evening lecturer of Computer Science at Northeastern University in Boston.

He became interested in politics and served for 25 years as an elected member of the Framingham Legislature. One interesting task, while being a member of the Legislature, was computerizing the Police and Fire Departments in Framingham.

On a district-wide political level, Joe became a member of Congressman Robert Drinan's strategy committee and, later, assisted with Barney Frank's first run for Framingham's congressional seat.

Joe's further involvement in the Town of Framingham was serving for 15 years as a Trustee on the Board of Framingham Union Hospital (the hospital famous for the Heart Study).

Life as a retiree

Joe and Linda now live part of the year on Hilton Head Island, South Carolina and part of the year in West Palm Beach, Florida, with trips back to Massachusetts during the summer where their four children and eight grandchildren live.

Joe and Linda Cooper in 2018

Jack Hirschfield – From Republic Aviation to Alarm Systems

Jack started his professional career as a physicist/scientific programmer at Republic Aviation on Long Island. After five years, and when Republic merged with Fairchild Hiller, he lost interest and looked for employment elsewhere. He applied for a programming job at Airborne Instruments Laboratory and was hired (initially unbeknownst to him) for the SR-71 project.

While waiting for his indoctrination into the SR-71 project, Jack filled out the necessary forms for his Top-Secret clearance. The security clearance agents conducted many interviews with people from Jack's past, even going to the apartment house where he had lived from the time he was four until he was twenty-five. They interviewed a number of "old ladies" who said that Jack "had been a good boy." When the Top-Secret clearance came through, Jack was unable to tell Ronni, his wife, anything about the nature of his work.

Jack was assigned to the "Ground Track" segment of the software project. Working with Dr. Richard Springer as an analyst, he wrote the code that computed the SR-71 mission flight path as it could be plotted on a large flatbed plotter. Jack recalled that John Altson had given him some guidance on how to make his software more efficient, relying more on keeping the software in the computer's memory and less on the use of magnetic tape drives. When the project started in 1965, there were no disks and no mass storage other than magnetic tapes.

The Ground Track software, which always tried to optimize the flight path to include as many targets as possible while missing the "avoidance areas," was extremely compute-intensive. With certain test missions, the software would crunch for hours, bringing the CDC 3200 "to its knees" and causing an occasional memory parity error.

Jack recalled there were twenty to thirty professionals on the staff: programmers, analysts, and managers. He loved the project and his getting an occasional glimpse of the SR-71 in its hangar at Beale Air Force Base. Although he left the project in 1968 when the staff was winding down, he continued to watch the SR-71's history and was proud to tell his friends (when he was allowed to do so) about his involvement in the project. He seemed to recall that the scope of the software encompassed one quarter of a million "IBM cards" of program source code.

Jack's first memorable trip to Beale Air Force Base in Marysville

The first trip to Beale was in the summer of 1966. Jack was given a choice of driving cross-country or flying; he chose driving. The first segment of the ride was with Ronni and her mother. Ronni, being a teacher, was off for the summer. When they got to Chicago, they picked up Jack's mother and continued to California. Along the way, Jack and family stopped at a number of national parks and found the whole trip "a great experience."

When they arrived in Sacramento, the computer system was not quite ready, so Jack was given another week of "vacation" which extended the group's touring into the states of Washington and Oregon.

Finally, Jack got to work on the software. Ronni stayed, and the two mothers returned home. On the weekends, they got to tour Old Sacramento, San Francisco, Mount Lassen, Carmel, Monterrey, Muir Woods and Yosemite. Some of his weekend trips were with Ronni and some with his AIL colleagues. Jack remembered that a group of his colleagues rented a cabin in Yosemite and awoke the next morning to find a herd of deer at the cabin's front door. All of these trips, according to Jack, were "most enjoyable" and resulted in very important team bonding.

The second summer at Beale

Ronni made the trip again with Jack; she returned after the summer to her teaching job while Jack stayed to "work the night shift."

Jack recalled being part of an AIL group which, on one particular weekend, drove all the way to Tijuana. They were crossing the border back into the United States when "carefree Jack" told the border guards that they had some liquor (true) and some LSD (not true). Of course, they were all led into a room and interrogated, being threatened with "strip searches." Jack confessed that he was joking, showed his Beale ID tag, and convinced the guards that they were good government contractors. The group was sent home and advised, "never to do that again."

Jack's professional life after AIL

Jack got involved in composition software, leading the effort to convert the South Bend Tribune newspaper from hot metal to cold type. He then formed a software company, Computer Interactions, which went public with Charles Plohn. They tried to develop computer schools but hit the market too early to be successful.

After this, Jack joined a friend's burglar and fire alarm company and spent the next twenty years perfecting the systems and selling them across the United States. One variant of the system was sold to IBM in order to centrally monitor many IBM buildings.

The company did so well that Jack was able to retire at age fifty-five, but retirement did not sit well with him. Jack's son suggested he come out of retirement, so Jack went onto his next project: aid the raising of venture capital money to create a GPS-based system to immediately detect vehicle theft. The system worked like this: If a vehicle was started without the proper deployment of a key-fob, the GPS monitoring would track the vehicle's motion and instantaneously send a signal to a centralized monitor system which would act on the theft. There were quite a number of innovations in the system: it could open car doors, communicate by voice commands, monitor air bags, etc. The competition was Lojack and, with Lojack, there was no way to report a theft until the theft was actually noted by the owner.

Regrettably, this product was also premature and too labor-intensive to install. They sold a few systems to luxury car owners but had no luck with fleet sales, taxi companies or trucks. Eventually, they ran out of money and the system was sold to a large automobile manufacturer. It is now known as "OnStar."

Ronni and Jack, 2017

Their grandson at Edwards in 2017

Joel Raphael – His First Job out of Graduate School

Joel started his assignment on the SR-71 project when he was fresh out of grad school in 1964. As a recent engineering school graduate, he was impressed to be working on a Top-Secret project at a time when we were dealing with things such as the Cuban missile crisis and the recent assassination of JFK. He could not remember what his title or job description was. He just remembered writing code for some reconnaissance system involving cameras and ELINT and learning about navigation and great circles.

Edwards Air Force Base, in Joel's words

"I was fresh out of graduate school and recently married, and this company sent me and my wife, along with a number of other couples and singles, off to some God-forsaken piece of desert where we couldn't tell anyone what we were doing. So, we told them we worked for a sewing machine company. And every morning we drove from the Antelope Valley Inn (which is still in Lancaster, California), and worked in air-conditioned trailers. If we had to walk from the trailer to a building, it was blazing hot! We kept frozen Milky Way bars in the trailer. Our wives stayed in the motel, but we left a car or two with them so they weren't stranded. On weekends, a bunch of us drove to L.A. or even Las Vegas. For Lenore and me, it was eye opening; we had been brought up in The Bronx and this was just not like the Bronx. We had little concept of what Lancaster would be like.

And then Joel went to Beale Air Force Base

"By the time we were in Sacramento, our first son, Mark, was 2-½ months old. So, it was a very different experience than Lancaster in that way. But again, every morning we carpooled out to Beale and came back at night to an apartment complex where a group of us lived. And on weekends, we went to Reno or Tahoe or sometimes to San Francisco. Workwise, it was similar to Edwards. We worked at our software, testing, showing progress. On one memorable day, Mel Berger and I were in the elevator that was associated with a hangar, a huge, hardened building. As the elevator was moving, it suddenly stopped between floors and the doors started to shake and move. There was an earthquake and the whole building was shaking. But because of the thick concrete wall construction, it was not in danger of collapsing. Eventually, the quake stopped, the elevator came to ground level, and the doors opened!

"I remember that as we approached the day we were supposed to be complete and demonstrate for some Air Force bigwigs, we were not quite ready. So, we set up the program to print out results as if it had completed the tasks. And I vaguely recall that we also had the system play some military music and even print out an American flag when it came to that point."

And his trips to SAC HQ in Omaha

"First of all, I did not serve in the military because I had a student deferment, and then I was married. So, meeting all of these high ranking military people was an experience for this introverted Jewish kid from the Bronx. I remember going to the Officer's Club for a meal and seeing a "kosher corned beef sandwich" on the menu and ordering it. And out came a corned beef sandwich with lettuce, tomato and mayonnaise. I also remember going to some famous restaurant in Omaha and having a prime rib dinner for under $5.00."

And after he left AIL in 1967

"When I was in graduate school, my Professor wanted me to go on for my PhD. When I said no, he advised me to go to work for Bell Labs because it was the closest thing to academia. When I graduated, Bell Labs did not hire me. But after 2-½ years at AIL, one of my dear friends, who worked at Bell Labs, arranged an interview and I got a job there. So, I left AIL. After five years at Bell Labs, they sent me on a two-year "rotational assignment" to the parent corporation, AT&T. The two years lasted for 19 years. I never went back to Bell Labs as an employee. I worked my way up at AT&T and eventually became the Director of Market Research for AT&T. I held that job for seven years. In 1991, I "retired" from AT&T, as they made me pension-eligible and paid me to leave. I then went into the Market Research business as an employee for two years and subsequently opened my own practice. After ten years, one of my AT&T colleagues was running the trade association for Yellow Pages and asked me to be the Director of the Research Institute for Yellow Pages. I did that for two years and then went back into my own business. Over the last five years, I have let that slide into another retirement because I travel the world with my wife, who is a renowned jazz pianist."

Lenore and Joel, 2017 Tribeca

The Summer of 1966 in Sacramento

Recollections of Three Families

From Linda Cooper

How accurate is anyone's 20/20 hindsight when the experiences took place in 1966? For me, Sacramento was such a large part of my young married life that many memories remain vivid, even after 51 years.

At the end of June 1966, Joe and I packed our suitcases, ourselves and our (then) three children in our car, left Commack, Long Island and started out on our big adventure to the West Coast. We traveled for 11 days and saw the wonders of this magnificent country along the way.

Joe had scoped out the Sacramento area well before our arrival so that by the time everyone came together on Marconi Blvd., the furnished apartments – purposely selected for their strategic locations around the pool – were ready for four families of the team. Added perks for every apartment were furniture, TVs (that I think had "rabbit ears") and weekly house cleaning services. Who could ask for anything more?

Funny – the most important thing we (or maybe someone else) did upon arrival that July 11th was to make sure everyone had toilet paper!

And strange – the actor, Ronald Reagan, wanted to be Governor of California.

And gruesome – a few days after our arrival in Sacramento, eight nurses were murdered in Chicago by Richard Speck.

John and Amy Altson, Jack and Ronni Hirschfield, Joel and Lenore Raphael, and Joe and I soon became good friends; the wives' relationship was especially close. Think about it — we had absolutely no commitments. The pool that was literally outside our apartments made it easy for us to spend our days there sunbathing, jumping into the pool to cool off; sunbathing, cooling off; sunbathing … etc. We were the original (outdoor) couch potatoes during July, August and September that year.

Oh yes, we did have one large commitment — grocery shopping. I was the one with the car and we all piled in and went shopping together once a week. Ho hum! Then back to the pool.

I didn't know what project Joe was working on. All I knew was that every day for three months he left our apartment and traveled a long distance to work. When he returned home at dinnertime, I had no idea where he had been, where he worked or what he was working on. He left in the morning and returned at dinnertime – just like any normal workday back in New York — except that he didn't talk about how his day had been.

In the evening, after the children were in bed, everyone came together around the pool to relax and socialize (it was important for the wives to have this additional relaxing time after the heavy-duty work of preparing dinner). We didn't need a baby sitter because the pool was right outside our apartment doors.

Usually, we all said goodnight around 10:00 PM, but two nights a week we left the pool early to congregate in my apartment to see the phenomenon taking place on television. It was called "Peyton Place." Oh, how we gaped at the 1966 risqué situations portrayed on the screen!

I thought California living was the best. It was so relaxed and casual. So carefree! It was pure luxury to go out for dinner or to a movie during the week – something we never did back in New York but something to which Californians never gave a second thought.

Who, in our young married lives, would ever think of spontaneously taking off and spending a weekend at Harrah's in Reno, Nevada? The casino called to us and although I don't recall what luck or lack thereof we had, I do remember enjoying Danny Thomas's performance in the nightclub.

When, up to that time in my life, had I ever experienced the meaning of the word "breathtaking" (the word "awesome" wasn't used in that context in 1966). Billowy clouds enfolded us as we reached the summit of the Golden Gate Bridge. I could swear we were driving right through them and heaven couldn't be too far away.

The weather in Sacramento was hot, hot, hot. Yet, the August weekend we spent in San Francisco with Jack and Ronni Hirschfield was cold, cold, cold, at least compared to the 100+ degrees we had left behind in Sacramento. Even though the temperature in San Francisco was around 60, it was raw and biting. Since it was late August, I assumed the fall season was setting in. What a surprise when we returned to Sacramento to find the weather was just as we had left it – still over 100 degrees late at night. Everyone was still around the pool; not too late for us to join them.

Being in California afforded us the opportunity to visit cousins of mine in Palo Alto and friends in Los Angeles, where we were exposed to the delights of the marvel called "Disneyland." One of the stops on our drive back to Sacramento along the Pacific Coast Highway was Sausalito. Where in Long Island was there anything like this little artist's colony that hugged the shore of San Francisco Bay? (Note: In my naiveté, I didn't even know such a thing as "the Hamptons" existed).

Being in California also opened my eyes to the diverse cultures of its different geographical areas. I noticed people native to the other side of the Pacific living in the northern part of the state, where we were. As we traveled south to visit our friends in L.A., the Hispanic culture was more in evidence. I noticed all this because it was not usual in Long Island to have a variety of cultures passing you on the street each day.

Miscellaneous personal remembrances: My six-year-old daughter taught herself how to swim in our Sacramento pool. She started first grade in a Sacramento elementary school. My four-year-old daughter celebrated her fifth birthday at the end of July and started Kindergarten in Sacramento in September. My three-year-old son just kind of went along with whatever was going on.

We left California at the beginning of October, just as the Jewish holiday of Rosh Hashanah began. Lenore and Joel Raphael celebrated with us; not with the traditional chicken soup and roast stuffed chicken dinner, because we had packed up and were heading home, but with a barbeque, of all things. Oh well, you do what you have to do, right? It was nice sharing the holiday with them.

It was hard to return to what I considered the more straight-laced way of living in New York. Once we were back into our old routine, the memory of our happy-go-lucky and laidback way of life in California faded. Not completely, though, because every now and then I remember wondering if it would be possible to replicate, in some way on Long Island, the lifestyle of those glorious months we spent in Sacramento, courtesy of Airborne Instruments Laboratory and the SR-71.

(Note: Even after the SR-71 was de-classified, Joe rarely spoke of his work on the project, other than in vague terms, and never mentioned the plane by name. Several years later, there was a news report about a plane crash (I think it was in Canada) where a politician was onboard and missing. A 'special" aircraft was deployed to search for him. As the broadcaster spoke about that aircraft, a sudden thought occurred me and I asked Joe if that was the plane he worked on in Sacramento. Sure enough, it was. And so, many years after 1966, I finally had an insight into the mysterious project that brought us to California).

Recently, I did some research on the SR-71 and must say that I was awestruck. To know that Joe was part of the team that assisted in its mission-planning fills me with pride."

From Joel Raphael

Yes, we remember that year for the High Holidays; we rented a couple of motel rooms near the synagogue so we could walk to services. And, as I recall, we had cut a deal to consolidate in one of the rooms on Yom Kippur so we could check out of one and pay to use the other until the holiday was over.

And speaking of the place where we lived, I arranged to rent a piano for Lenore because she couldn't be without one for three months! Who would have known that she would go on to play professionally?

Another story this reminded me of: On the first trip Joe and I and a couple others made to northern California before that summer, we decided that when we landed in San Francisco we would go buy presents for the wives and kids because if we waited for the way back, we would probably forget and then have to buy things in the airport. So, we landed, rented a car and went to the first town south of the airport, San Bruno, and we found a children's store: Todd's Toys, on the main street. While shopping, I knew that Lenore had a cousin who I had never met who owned some sort of store in San Bruno. So, I asked the cashier if, by any chance, she knew a storeowner named Rudy Solomon. She looked up and said, "Sure, he owns this store and he is in the back." So, I went back and introduced myself and, subsequently, when Lenore came out, she got to see him, as well. I might have guessed this was his store because his brother, whom I knew well, was named Todd.

And, Linda Cooper mentioned that Joe never discussed his work on the SR-71 (Blackbird). That was probably true about most of us. About ten years ago, my grandson, who was then about 15, came to visit us in New York and we took him to the Intrepid, the aircraft carrier used as a museum on 42nd Street in the Hudson River. They have a Blackbird sitting on the deck and he enjoyed getting into the cockpit. When he came out, he was one of the first people I ever told that I had worked on its mission control when his father was first born and that we used to tell people that we worked for a sewing machine company as a cover story.

From Jack Hirschfield

Since Ronni was a teacher, she had to return home after the summer was over. Meanwhile, I stayed in California programming away and, after five weeks of very little communication from me, she called on my birthday. "Hello honey, happy birthday," she said. I replied, "So what did you buy me?" She said, "I hope that you are sitting down, because I just bought you a car and if you do not come home soon, I am buying a house"

I recall going home the next weekend!

A brief history of Airborne Instruments Laboratory (AIL)

As extracted from the 1991 issues of the AIL Record: 'Brought to you by your host and the generosity of John Menechino.' (Key excerpts from website http://www.pauls-page.com/ailhistory.htm):

- 1942 AIL begins work on the AN/ASQ-2, a system which would automatically fire markers and retro-bombs for the location and destruction of submerged submarines. This is a refinement of the Magnetic Airborne Detector, AIL's first product.
- 1958 AIL is acquired by Cutler-Hammer.
- 1959 AIL is awarded the prime contract on the ADS-1 (later named the USD-7), the most sophisticated airborne system of its type ever ordered. Among the subcontractors to AIL was Sperry Gyroscope. Raytheon, and Sylvania.
- 1959 AIL receives an $8 million Air Force prime contract for 29 Video Integrating Groups (VIG) to filter radar data and eliminate interference.
- 1961 On June 24th, the AIL-developed, "Topside Sounder" satellite is launched. The satellite is designed to measure the degree to which the ionosphere reflects radio waves. The "Topside Sounder" reached a peak altitude of 633 miles.
- 1962 The USD-7 system undergoes its first flight test in Greenville, Texas. The USD-7 award of $38.9 million in 1959 was AIL's largest contract to date.
- 1963 The Department of Aviation Systems Research, working under a 3-year contract with the Federal Aviation Administration (FAA), develops a mathematical model that forms the basis for a computer program capable of predicting the maximum practical operating capacity of any airport.

- 1966 The AN/PPS-5 portable battlefield radar is being tested in Germany by NATO. While the PPS-5 was being transferred by ship in the North Sea, the ship's radar went dead in a heavy fog. The crew unpacked the AIL portable radar which worked perfectly and saved the day.
- 1969 AIL is awarded a $40.4 million contract for the AN/TPX-42 air traffic control system. It will provide 3-dimensional position data correlated on the controller's display with aircraft identity.
- 1977 Space Shuttle "Enterprise" goes on its first solo flight (separated from a 747) at the Dryden Flight Research Center. AIL designed the landing system for the shuffle.
- 1978 AIL signs a $17.8 contract to provide a Vessel Traffic Management System for the Suez Canal.
- 1980 Scientists and engineers from the Tokyo Astronomical Laboratory visit AIL for design review of low noise RF Amplifiers and converters AIL is providing for 2 radio telescopes. The $3 million contract is AIL's largest ever for radio astronomy equipment.

The SR-71 mission operational flow

I was never able to determine the exact date of when Airborne Instrument's MP^2 software was formally accepted by the Air Force and when the operational use began. An educated guess would be in 1967.

We know that MP^2 was deployed in three operational bases (for optimal geographic coverage of the desired target areas):

- Beale AFB, the home of the 9th Reconnaissance Wing (9 RW)
- RAF Mildenlhall in the U.K.
- Kadena Air Base in Okinawa

These three bases had both pre-mission and post-mission technical support: Pre-mission support involved running the MP^2 software (and other Air Force-created software) while post-mission support involved analyses of the data recorded on the mission.

As you might also suspect, the Pentagon and SAC Headquarters in Omaha were also involved.

A summary of the operations performed at Beale (Wikipedia)

On 15 October 1964, the Department of Defense announced that Beale would be the home of the new, supersonic reconnaissance aircraft, the SR-71 Blackbird. The provisional 4200th Strategic Reconnaissance Wing (4200 SRW) activated on 1 January 1965 in preparation for the realignment, and the new wing received its first aircraft, a T-38 Talon, on 8 July 1965. The first SR-71 did not arrive until 7 January 1966. The SR-71 was developed from the Lockheed A-12 reconnaissance aircraft in the 1960s for the Central Intelligence Agency by the Lockheed Skunk Works as a black project. During reconnaissance missions, the SR-71 operated at high speeds and altitudes to allow it to outrace threats; if a surface-to-air missile launch was detected, standard evasive action was simply to accelerate.

With the arrival of the SR-71, the strategic bombardment mission at Beale was phased down, being replaced by the Strategic Reconnaissance mission. The 9th Strategic Reconnaissance Wing was moved from Mountain Home AFB, Idaho and reassigned to Beale on 25 June 1966. The 9th RW has been the host unit at Beale to the current day.

With the activation of the 9th SRW at Beale, the wing absorbed the assets of the provisional 4200th SW. This allowed it to stay with the 14th Strategic Aerospace Division. The wing performed strategic reconnaissance in Southeast Asia beginning in 1968, frequently deploying the SR-71 to Kadena Air Base, Okinawa where it operated over areas of the Pacific and Asia. The wing provided photographic intelligence for the Son Tay prison camp raid named Operation Ivory Coast, in North Vietnam, November 1970. After the Vietnam War, the SR-71 established a level-flight-at-altitude record at 85,131 feet and a straight-course speed record of 2,194 mph.

On 1 July 1976, the U-2 joined the SR-71 in the 9th Strategic Reconnaissance Wing, giving the unit two of the most unusual aircraft in the world. The "Dragon Lady" had gained national and international recognition with flights over the Soviet Union, China, Cuba, and Southeast Asia. The U-2 was the perfect complement to the SR-71. The Blackbird was designed to capture time-sensitive intelligence, especially in denied airspace. Whereas the SR-71 was designed for rapid infiltration and exfiltration, Dragon Lady was designed to loiter in its area of responsibility and continuously collect while in the air.

The SR-71 mission was closed on 1 January 1990. Although it made a brief revival in the mid-1990s, today the aircraft remains retired.

A summary of the Mildenhall operations (Wikipedia)

With the arrival of the 306th SW, Mildenhall also became known as SAC's European Reconnaissance centre. For many years, various types of Boeing RC-135 reconnaissance aircraft were observed regularly arriving and departing from the Mildenhall runway. Most of these aircraft came from the 55th Strategic Reconnaissance Wing at Offutt AFB, Nebraska, and had the capability to receive radar and radio signals from far behind the borders of the Communist Eastern Bloc. From Mildenhall, the RC-135s flew ELINT and COMINT missions along the borders of Poland, the Soviet Union and Czechoslovakia. The twenty or so specialists on board the RC-135s during such missions listened to and recorded military radio frequencies and communications.

The next significant event in Mildenhall's history came with the arrival of Detachment 4, 9th Strategic Reconnaissance Wing in 1976, which controlled rotational Lockheed U-2 and SR-71 Blackbird operations from the base. It is not known when SAC first began making reconnaissance flights in Europe with these aircraft. There are indications that these fast aircraft have been operating in Europe since the end of the 1960s, with an SR-71 making a stopover in August 1970 at RAF Upper Heyford, England, before a mission over the Middle East.

These aircraft carried out strategic photoreconnaissance missions for NATO and the USAF within the framework of the SALT I Agreement of 1972. Under this agreement, the Soviet Union and the United States reached agreement on a partial freeze on the number of offensive nuclear weapons, and these flights were to check that the Soviets were adhering to the agreement.

As well as the photo missions, the 9th SRW gathered telemetry signals from Soviet missile systems. Such missions were carried out using the SR-71 and U-2/TR-1 aircraft, and Boeing RC-135s from the 55th SRW. This information was analysed, together with information originating from reconnaissance satellites, to present an intelligence picture for analysis to assemble a good picture of Soviet activities for national decision-making.

From their arrival, until the departure of the last SR-71 on 18 January 1990, the 306th Strategic Wing's SR-71 and U-2 aircraft came to symbolise RAF Mildenhall in the local public's eye.

A summary of the Okinawa operations and the "Habu" nickname (Wikipedia)

The SR-71s averaged approximately one sortie a week for nearly two years. By 1970, the SR-71s were averaging two sorties per week. By 1972, the Blackbird was flying nearly one sortie every day. ***While deployed on Okinawa, the SR-71s and their aircrew members gained the nickname Habu (as did the A-12s preceding them) after a Southeast Asian pit viper that the Okinawans thought the plane resembled.***

Habu, the snake

Who planned the missions?

This process was articulated well in Colonel Richard Graham's book, *SR-71 Revealed – The Inside Story*. The following is an excerpt from that book:

> "The highly classified National Reconnaissance Office (NRO) was created in the early 1960s to combine satellite reconnaissance projects underway at the CIA and Defense Department. The office was headed by the Assistant Secretary of the Air Force. The existence of the NRO was an official secret for many years, until its name and overall mission were made public in 1992. Allocation of NRO assets, including the SR-71, was controlled by the Director of Central Intelligence (DCI). Any user that needed intelligence acquired by NRO resources had to make a request through the DCI's office. A committee then reviewed intelligence requests daily, prioritized them, and directed the NRO to position satellites or aircraft as necessary. The SR-71 was part of this process and, consequently, was considered a national reconnaissance asset.
>
> SR-71 operational sorties were flown under the rules of the Peacetime Aerial Reconnaissance Program, or PARPRO (pronounced "Par-Pro"). PARPRO was created for all military reconnaissance assets (air-breathing aircraft) and was managed at the Joint Reconnaissance Center (JRC) in the Pentagon. Every peacetime reconnaissance mission flown worldwide was continuously monitored from start to finish at the JRC, 24-hours a day (called "the watch").

The JRC was further responsible for determining the threat to our aircraft on each mission and obtained the appropriate approval authority to fly each of them. If the mission involved overflight of foreign countries, it was JRC's job to obtain the necessary permission from the State Department. Very few people in the Pentagon were knowledgeable on the SR-71's capabilities or had a SENIOR CROWN security clearance. For those reasons, a former Habu was usually assigned to the JRC and performed duty 'on the watch,' able to answer questions if a mission did not go smoothly."

What, then, was an "operational mission"?

Again quoting from Colonel Richard Graham's book:

> *"Operational Missions:" "Operational" missions were those that flew through what was called a "sensitive area." While gathering intelligence on foreign countries, we flew in airspace that was defined as a sensitive area. There we could expect any kind of reaction from a foreign country and had to be alert for any possibility. Flying in the sensitive area, all of our airborne equipment, systems, and navigation had to be 100 percent perfect, otherwise it was an automatic abort.*

> *When Don and I flew through a sensitive area, cockpit communications were minimal. We said only what was necessary and paid strict attention to our job at hand. Your sense of awareness became keenly intensified as you wondered what you might see today. We logged our operational missions separate from all other types of flying activity, and after flying around 50 operational missions, Habus were awarded an Air Force Air Medal.*

> *Being under command of the former SAC, they also wanted a final vote on each operational mission and, consequently, had to approve each one of them. Located at SAC Headquarters in Omaha, Nebraska, was another agency with functions similar to the JRC, called the Strategic Reconnaissance Center (SRC). The SRC had a staff of officers, who for the most part, were former crewmembers in their respective reconnaissance aircraft, SR-71, U-2, and RC-13.5.*

SRC had to send an approval order, called the "Execution" message, for every scheduled operational mission. We could not fly without the "Execute" message from SRC. The "Execute" message was transmitted via secure communications directly to each Detachment Commander and included numerous other addressees who needed to know the mission was a "go." Once the Detachment received the "Execute" message it was up to them to make it work.

The majority of our operational sorties were flown repeatedly, with only minor changes to the track or sensor operations. Most of these "routine" missions were flown to gather what was called "Indications and Warnings" (I & W) intelligence on other countries. Flying over the same sensitive areas on a regular basis allowed intelligence analysts to determine such things as troop movements, changes to the EOB, and aircraft deployments—all good indications and warnings that something was about to happen. If the I & W intelligence indicated a high level of activity, we could then focus our intelligence gathering in greater detail on a particular geographic region."

Later in this book are the details of the interview with Colonel Frank Stampf. Frank served as the SRC at SAC Headquarters for a while and he indicated that the SRC would select the optimum operational base for the mission. Frank said that it was "not always obvious," because flights frequently went over the Arctic Circle.

Frank also said that "high priority" missions were exceptions to the normal operational flow. In these instances, the requests for reconnaissance targets would emanate from one of the geographic "theatres." An example of this might be when CINCPAC (the Navy) wanted reconnaissance of the Russian missile submarine fleet near Murmansk and far above the Arctic Circle. CINCPAC needed to be constantly aware of these submarine maneuvers.

So, was the SR-71 "spying"?

Technically, no. According to the *Free Dictionary*:

> **spy** (spī)
>
> n. pl. spies (spīz)
>
> 1. One who secretly collects information concerning the enemies of a government or group.
>
> 2. One who secretly collects information for a business about one or more of its competitors.
>
> 3. One who secretly keeps watch on another or others.
>
> v. spied (spīd), spy·ing, spies (spīz)
>
> v.tr.
>
> 1. To watch or observe secretly: was sent to spy out the enemy camp.
>
> 2. To discover by close observation: "[They] are continually prowling about on all three decks, eager to spy out iniquities" (Herman Melville).
>
> 3. To catch sight of; see: spied the ship on the horizon.
>
> v.intr.
>
> 1. To engage in espionage.
>
> 2. To investigate or observe something, especially in secret: spying into the neighbor's activities.

The SR-71 was clearly marked with the American flag. The crew wore the American flag on their pressure suits. The SR-71 flew in international airspace. There was no subterfuge; the missions were, according to Colonel Rich Graham, "keeping a watchful vigilance."

PARPRO missions were flown in the international airspace near China, Cuba, Iran, Iraq, Libya, Nicaragua, North Korea, Persian Gulf region, Soviet Union, Vietnam, and Warsaw Pact countries.

Operational support of the missions

The Beale support was conducted out of the SAGE Building, while the Mildenhall and Kadena sites had "Mobile Processing Centers" or MPCs.

Courtesy of Paul Crickmore, here is a picture of some MPCs:

One of the vans shown in the picture above housed the CDC 3200 used for SR-71 mission-planning, programming the aircraft's sensors, and providing an initial scan of the ELINT collection.

Pre-mission

The pre-mission support was similar in all three operational bases.

First, in the words of Colonel Rich Graham (reprinted from *SR-71 Revealed*), the planning perspective from a pilot's point of view:

> *"The SR-71 Mission-planning Branch developed all our missions. For every new mission, they had to do what was called a 'feasibility study.' This process entailed looking at the area of interest to insure it could be covered adequately by the SR, the intended route of flight, foreign country overflight considerations, placement of air refuelling tracks, and where best to originate the mission. Once it proved feasible and the OK was given to develop the mission, planners went into high gear working out the details. Specifics such as: what suite of sensors to use, where and when each sensor should be ON/OFF, where to best utilize the air refuelling tracks, fuel requirements, where the JP-7 was coming from, timing of the mission, etc., etc., all had to be planned out in excruciating detail.*
>
> *Along with the mission tape for the ANS, the planning branch also developed the aircrew's computer-generated flight plan. The RSO's flight plan was much more detailed than the pilot's, since it included data telling him when specific sensors were ON/OFF and where his navigational fix points were located. Mission planners drew the route out on maps and pasted computer-generated labels along the track to show specifics such as elapsed time, fuel remaining, Mach, altitude, heading, bank angles, coordinates, and distance to the next turn point. Habus referred to the line drawn on our maps, depicting the route of flight, as the 'Black Line.' After the mission maps were made, 35mm pictures were taken of the entire route and developed into a roll of film, to be placed in each cockpit's moving map projectors.*

The pilot's map projector was on the forward instrument panel and sat right between his knees. The 35mm mission film was loaded before flight and preflighted by the mobile crew. After the pilot turned his projector ON he had a 4 1/4" x 4 1/4" moving map display of the entire route with all its labels. Placing the map projector switch in 'AUTO' allowed the film to move over the projector screen in concert with the aircraft's true airspeed. A spring-loaded, fingertip-slewing switch in both cockpits allowed us to fast forward or reverse the film wherever we wanted, in a matter of seconds. Once Habus became familiar with all the mission materials, they found they could fly the entire mission with reference to the map projector alone.

The map projector's 35mm film always started with the title frame, displaying the specific mission number we were flying and date of revision. During preflight, the mobile crew checked the mission number and date against the flight plan and ANS to insure all three had the same mission number and date of revision. Would you believe the next frame was usually a photo taken from a Playboy or Penthouse magazine, to keep the crew company at 80,000 feet? The mission planners enjoyed surprising us with a different photo each time. The RSO's map projector consisted of a 9 x 9-inch glass projection screen giving him a larger presentation and greater detail than the pilots."

I conducted three interviews with Air Force personnel involved in pre-mission operational support: Frank Huddleston, Kurt Pfannuch, and Tony Shelton. The results of these interviews are in later chapters.

Here, however, is a bulleted list containing a summary, from their viewpoints, of the pre-mission support:

- The mission route was codified in a deck of IBM cards; the targets were provided as a printout. "Route Search" software may have been used to utilize a library of "canned" pre-designated routes.
- The coordinates of the targets were plotted on a huge wall map and the curves drawn for the SR-71 turning points.
- All of this was coded on IBM cards and fed into the mission-planning software on the Control Data 3200 staffed by computer operators.
- Targets were prioritized, and the mission was planned so as to hit the high priority targets and as many of the secondary targets as possible.
- If there were error messages coming out of the CDC 3200 software, the targets' coordinates would be altered until there were no remaining errors and there was a "clean mission."
- The mission-planning input deck would be re-run to show what targets were covered. This deck would also produce a Mylar tape that would turn the sensors on and off while flying the SR-71 in autopilot mode.
- Labels were created for the strip chart used by the RSO (some sensors needed to be turned on and off manually), and a checklist had to be strapped to the RSO's legs in the event that the filmstrip jammed.
- Software was run to calculate the optimum fuel load for each SR-71 fuel tank.

Post-mission analysis

The Mission Recorder System (MRS), from a pilot's perspective (quote from Colonel Rich Graham)

> *"For ease of maintenance, the aircraft was equipped with a Mission Recorder System (MRS) that recorded every three seconds on magnetic tape, various parameters of the aircraft and its associated systems (engine, hydraulic, DAFICS, electrical, ANS, etc.). The MRS also recorded signals from various aircraft data sources, including analog transducer sources, digital information sources, cockpit and external voice communications, DEF systems, and event information sources. After flight, maintenance removed the MRS tape and, through a computer process, converted the electrical inputs into meaningful information. The MRS maintenance shop distributed the recorded information to individual maintenance shops for experts to look over and see if anything was going wrong with their particular system. When crews had difficulty describing maintenance problems on specific aircraft systems, such as the inlets or engines, they occasionally went over the MRS data, along with maintenance, to describe what they saw or felt in the cockpit during the time(s) in question."*

The Air Force's post mission team was interviewed for this book; the chapters of Tony Shelton and Frank Huddleston reflect their operations that included, by definition, the extraction of the MRS data from the SR-71.

A quick summary of the post-mission processes

- Pulling the telemetry data from the sensors' recordings.
- Running the Photo Program that generated the geographic coordinates of each photo for later analyses.
- ELINT processing – recording the location and signal strengths of enemy radar.
- A printed custom-designed report for every system on the airplane – environmental systems, engines, inlets, fuel, cameras, radar, defensive systems, etc. Reports were picked up by someone from the MRS shop and distributed to the different shops on the flight line – usually three to four hours after landing. The defensive systems analysts in the headquarters building got a report, as well, as did the DCO, who got a sonic boom track with a position reported every two minutes while the aircraft was supersonic.

Interviews with Air Force Operations

The support provided by Air Force personnel before and after a mission was instrumental in the success of the SR-71 program.

I was lucky enough to locate three individuals who provided that critical support. Following are their stories. I should also add that two Air Force operations personnel declined to do an interview; they were concerned that some aspects of their work might still be classified.

Frank Huddleston, Mission Planner

Frank was only eighteen when, in 1975, he received his assignment to do mission-planning for the SR-71. He did not know quite what to expect of the job, but he did it for almost three years. It turned out to be the most enjoyable job of his career and, according to Frank, he could have gone on doing the job for thirty years.

The days at Beale

Frank's assignment started at Beale Air Force Base in Marysville, California, as part of the Beale Reconnaissance Wing. He described his workflow in the following way:

1. SAC Headquarters (Omaha) would provide the proposed flight route and the proposed targets through Beale Operations. On occasions, SAC might just provide the targets and Operations would supply the route. In any event, the mission route was codified in a deck of IBM cards; the targets were provided as a printout.
2. Frank would plot the coordinates of the targets on a huge wall map. He would draw the curves for the SR-71 turning points and indicate where to turn the sensors on and off. All of this was coded on IBM cards and fed into the mission-planning software on the Control Data 3200 run by staffed computer operators. (While the 3200 software computed the turning points for the curves, there was no way to visualize these curves until Frank plotted them on the wall map.)
3. Targets were prioritized, and Frank would plan the mission to hit the high priority targets and as many of the secondary targets as possible.
4. If error messages came out of the software, Frank would continue to alter the targets' coordinates until there were no remaining errors and he had a "clean mission."

5. Frank would then re-run the mission-planning input deck to show what targets were covered. This deck would also produce a Mylar tape that would turn the sensors on and off while flying the SR-71 in autopilot mode. (Frank said that the job of the Reconnaissance Systems Operator (RSO) was more to monitor the sensors, rather than switch them on and off (the job of the mission's Mylar tape).
6. An Operations person made a 35 MM filmstrip of the route for the RSO.
7. When the mission was over, Frank would check to see that the sensors properly covered the targets.

Many of the SR-71 training missions were flown out of Beale and all of the training missions that Frank planned were flown out of Beale.

Frank's comments on the wall map construction

"The "wall maps" were navigational charts. The ones I recall using were (GNC) Global Navigational Charts and (JNC) Joint Navigational Charts. They were large maps, about 4-5 feet wide, kept in a bank of wide cubby holes, stored flat, and labeled.

"I used a T-square, simple compass and dividers, a Weems plotter, a couple of pencils, and felt tips to draw the flight path. Then, in the target area, I would pencil in a line showing the outer reach of the cameras and SLR. The goal was to photo or get SLR images of as much of the area between the flight path and pencil line. The cameras were limited by the amount of film, film speed and range. I used the TEOC camera for the priority targets, backed up by the OBC cameras as well as the tracking camera. The TEOC cameras were target-specific and the others were area cameras. If I could get images of the target by all the cameras AND the SLR, that was a bonus.

"During time on, the SR never overflew any non-friendly company in the Asia/Pacific/Vladovostok region, only flights in international airspace. The SR altitude allowed us to look 'over the fence' far into a country. I planned several Cuban missions as well, until President Carter ordered them stopped."

Three trips to Okinawa

Frank was asked to do mission-planning out of Okinawa on three separate occasions, starting with when Saigon fell to the North Vietnamese. The first deployment of spring 1975, assigned to the 9th RTS, his duties were as an ELINT analyst processing electronic intelligence data from operational missions in response to the collapse of South Vietnam. The remaining two deployments were as an operational mission planner with the 9th SRW/Intel Branch.

It was very hot in Okinawa and the computer, housed in Mobile Processing Centers (MPCs, or 20-foot trailers), would frequently overheat and not perform well. During Frank's tours, the Okinawa base was the operation center for SR-71 reconnaissance flights of Russia, China, Vietnam, and North Korea. The missions that Frank planned were all flown on the periphery of the target nations, offshore and in international airspace. Frank planned several Cuban missions that were over-flights of the island of Cuba.

Frank's observations while on the job

The computer in Okinawa wasn't the only thing to get overheated. Frank, himself, got frustrated when he had to rely on computer operators to run his jobs and they were focused on running the day-to-day operational software, not his high-priority mission deck. Out of necessity, Frank learned to run the 3200 software himself and did so occasionally.

Missions were flown based on the weather and the target opportunities. Sometimes there would be as many as three missions on a given day; sometimes there would be a three-day break between missions. On the average, one mission was flown per day. The performance of the side-looking radar (SLR) was affected by the weather.

There were some MPCs at Beale, but most of the computer runs were done in a modern computer room with raised floor and air conditioning.

On his last day on the job, Frank wanted to take home some photos of images taken by the SR-71 while on U.S. training flights. He was not allowed to do so because, at the time, these photos were sensitive, as they indicated the RS-71's imaging capabilities. Too bad, as there were some wonderful shots of the Golden Gate Bridge!

In Frank's own words ... leading up to the mission-planning job

"I was there 1974-1978. Refueling was planned in operations, which shared the 3rd floor with Intel. They were mostly officers, in ops 9SRW/DO, with the graphics guy that did the mission filmstrips. And all from memory if details aren't precise. I was an ELINT analyst assigned to the 9RTS working the ELINT data from the SR and U2s. I then was reassigned to 9SRW/IN to do mission-planning for the SR. They used a few ELINT and Photo Interpreter guys to do this and we would rotate to the DETs. I was exclusive to DET 1, Kadena.

"I did the Intel mission-planning – that is target selection along the route, sensor operation, post-mission ELINT analysis, 1974-1978. The software and computer I used was the enormous Control Data machine located in the 9RTS. I turned on cameras, ELINT and SLR using individually punched IBM cards. There was a whole other group, Operations, that laid out the mission and refueling. The mission came from SAC/Offutt with the route laid out and requests for specific targets.

"My Air Force specialty was Electronic Intelligence Operations Intercept/Analyst, Air Force Specialty Code 205X0, or "205" for short. I was trained to observe, document, and analyze electronic emissions (radar signals) for use in devising electronic countermeasures. I was assigned as a new young Airman to the 9th Recon Technical Squadron (9RTS) at Beale Air Force Base, California, under the 9th Strategic Recon Wing. Their mission was primarily the operation and support of the SR-71 and later, in 1976, the U-2 aircraft.

"In the 9th RTS, I had two assignments. 1) Listening to U-2 radar threat tapes taken from U-2 missions flown in the Mideast and 2) Analyzing electronic emissions from current SR-71 reconnaissance missions flown in the Western Pacific (Vietnam, Korea, China and USSR). U-2 tape analysis consisted of listening to audible signals and watching signals on an oscilloscope, looking for aircraft threats, primarily SA-2 surface-to-air missile radars (acquisition, targeting, and missile launch signals). I did not do this for very long; a few months, perhaps. My section was located in a windowless concrete building on the 2nd floor next to the computer room. We operated behind a cipher-locked vault door. In our section, we had two shifts of 4-8 analysts. There was also a small civilian team of various industry representatives from CDC, Adage and others that I can't recall.

"At this time, January to about March 1975, I was moved to the other assignment of analyzing SR-71 ELINT (electronic intelligence) using data from the aircraft mission tapes. They did this in order to train me rapidly on the process so that I could replace senior analysts who were being deployed to DET 1 (Detachment 1, OL-KA, Kadena AB, Japan (Okinawa) as the fall of South Vietnam was evident. The 9th SRW activated the MPC (Mobile Processing Center) at DET 1 to process intelligence collected as soon as it was retrieved from the aircraft. Shortly after the fall, I deployed with a Senior Analyst and processed mission ELINT data. Normally, the data is returned to Beale AFB and the 9th RTS where it would be processed. Analysis of the SR-71 ELINT data then was done by utilizing something high-tech at the time (late 60's-early 70's) – a computer graphics terminal (Adage Graphics Terminal). It was a computer about the size of a large sofa, with a CRT screen, light pen, and ball roller (mouse equivalent).

"Processing data on the Adage Graphics terminal was done by using a series of commands and setting parameters to identify the signal frequency, bandwidth, PRF (pulse recurrence frequency), etc. The analyst had a large printout of the mission flight, location, and signals intercepted. Using the printout, the analyst would scan the numbers for specific parameters previously mentioned and use the printout with the Adage to process the mission.

"Simply explained, once the parameters were set, a display appeared on the screen (black face with white characters and images), appearing like dropped spaghetti. Where the "spaghetti" lines intersected was probably where the emitter was located. Using the light pen and ball roller, the screen could be "cleaned up" and a circle or oval would be placed around the points of intersection. This circle could be, literally, a mile in diameter, and the smaller the circle, the closer the intersection of lines, indicated to the analyst a better location of the emitter. The data would be compared manually to a paper EOB (Electronic Order of Battle) and an attempt made to determine if an already identified radar emitter was identified. After completion, I believe the results were forwarded to the SRC, (Strategic Reconnaissance Center, SAC, at Offutt AFB).

"I did this for one two-month deployment to DET 1 and continued on this assignment for about a year. One day, a Major and a Technical Sergeant came to meet with me after meeting with my supervisor and officer-in-charge. They were from the 9th SRW Intel Mission-planning section and were recruiting/selecting an ELINT analyst (AFSC 205) to move into a Mission Planner position in the Wing. I was very young, 19 as I recall, married with a small child. Looking back, it was a surreal time of my life.

"But for some reason that is still unclear, either I was "volunteered" for the assignment by my supervisors or selected by the Wing to move to the new assignment. I was adamant that I did not want to go. I was comfortable with friends I had attended Tech School with and worked with the whole time and I was feeling very comfortable in my assignment. I went with the Major and Sergeant to the 3rd floor of the same building, behind another cipher lock vault door. They led me into a big open room with 15-foot ceilings and several long (15-20 feet) wooden tables with varnished tops in the center. Along the walls of the room, were about six desks with senior NCOs or officers at them in blue 1550 uniforms, quietly watching…me.

"The Sergeant and Major that brought me there began to explain their mission and what my duties would be. When I explained that I appreciated the opportunity but didn't want to leave, the Sergeant, who would be my supervisor, just smiled and calmly asked me if I thought I had a choice. He turned out to be a great supervisor, leader, and friend. A calm, intelligent and very funny guy, he made this job one of the best, if not the best, jobs I ever had."

Kurt Pfannkuch – After many tours of duty, his SR-71 home at Beale AFB

Kurt joined the Air Force in 1962 and had several tours of duty as a B-52 navigator-bombardier supporting Operation Arc Light over South Vietnam. His assignment kept him deployed from multiple Air Force bases, traveling a great deal on his assignments. One problem was that they kept closing the bases from which Kurt was deployed!

Finally, in 1969, someone at Strategic Air Command headquarters in Omaha said that they needed help in the computer areas. It was more efficient to train an air operations person on computers than to train a "computer nerd" on air operations, so they recruited Kurt, gave him three months of computer training, and put him to work for two and a half years working on war plans – specifically laying down the Air Force nuclear war attrition plan. Given the areas of threat and the current Air Force deployment, Kurt computed how to assess planned strike routes against the effectiveness of enemy defenses.

The Vietnam War was escalating and, after his three-year assignment at SAC HQ, he was under consideration for reassignment to the active crew force. Because of Kurt's previous tours of duty, however, he was dropped from consideration. Instead, he found out about an opportunity in the SR program at Beale and "jumped on it." He was there from 1970 until 1984, when he retired from the Air Force.

Pre-Mission-planning at Beale

There were four distinct categories of pre-mission-planning: (1) Running the mission-planning software that defined the mission route; (2) Comparing that route to a database of potential targets of interest; (3) Creating the sensor on/off points based on the inputs of Intelligence planners; (4) Running programs to create supplementary pre-mission support such as labels for the strip chart used by the RSO (some sensors needed to be turned on and off manually), a checklist to be strapped to the RSO's legs in the event that the filmstrip jammed; and the Mylar tape that fed the (ANS) navigation system.

The history of the mission-planning software

The mission-planning software developed by Airborne Instruments Laboratory was written in the FORTRAN IV language. The Air Force "pretty much kept it as Bible" and many of the staff were "in awe of it". Few changes were made, most of them to accommodate new sensors and new targets in the target database. Frank recalls, "It was amazing what you could do with that primitive equipment."

In 1978-1979, the Air Force acquired a Systems Engineering Lab (SEL) 32/55 computer and ported the FORTRAN IV code over to FORTRAN 77. It was still an (archaic) punched card and magnetic tape system, however, and the CDC 3200 logic was kept basically intact.

The Beale computer was housed on the second floor of the SAGE Building. It was one of several buildings developed by NORAD in the 50s to house what computer power and communications they had to control and coordinate regional air intercept operations. The three-foot thick concrete walls could withstand almost anything short of a direct nuclear hit. There were also two Mobile Processing Centers (MPCs). One saw service at multiple forward area locations such as RAF Mildenhall and one in Okinawa. Both MPCs were capable of the photographic processing required to support the missions.

The mission-planning software, in more detail

There were two aspects to the mission-planning software: (1) Flight Simulation to include takeoff, landing, fuel consumption, refueling points, etc. It always maintained the desired track. (2) Route Search, which pulled in targets from the target database and created a "swath" on either side of the track to include the selected targets. The mission-planning software then created a list of all of the targets that would be within the (swath) collection area.

Intel planners then, knowing the plane's sensor load, created the sensor on/off points from the Route Search printout. The sensor load typically consisted of optical cameras, SLR (side-looking radar), and TEOC cameras. The Mylar tape for the ANS navigation system was created at the end of this process.

And by the way...

The same mission-planning software was modified in the early days of NASA, planning the trajectory trail for spacecraft pre-entry. Interesting!

And there was a post-mission analysis group

The post-mission group was responsible for gathering the data from the sensors and preparing for in-depth analyses. A short list of their responsibilities: (1) Pulling the telemetry data from the sensors' recordings; (2) Running the Photo Program that generated the geographic coordinates of each photo for later analyses; (3) ELINT processing – recording the location and signal strengths of enemy radar.

For a detailed discussion of the post-mission maintenance system, please see Tony Shelton's chapter. The system was the forerunner of today's infamous black box and recorded aircraft data from over 200 collection points at 3-second intervals, providing the input data for extensive analysis of flight accuracy, subsystem operations, and projection of future maintenance problems or failure.

Kurt's career after the Air Force

Kurt stayed in Yuba City after his retirement, trying to get by on his Air Force retirement. That did not work economically, so he looked for employment locally. After a few years, he moved to Sacramento and worked for the California Department of Corrections doing software support. He finally retired in 2002.

Kurt's wife is from Sweden, so they go there each summer. He has two daughters and four grandchildren.

A funny anecdote…

There was one mission in which the radar system showed rows of suspicious objects. An excited young lieutenant suggested that they fly another mission to get more details, but a detailed analysis of the photographs taken revealed that they were nothing but large piles of dung – hardly the basis for a follow-on mission.

Tony Shelton – Kicking off the post-mission analyses

Tony Shelton was with the SR-71 program from 1974 until 1978. During this time, he worked in the Autoflight Programming Division under the Director of Operations (DO). His division worked on pre-mission flight facilitation and post-mission analysis; Tony was primarily involved in the post-mission analysis side. He was the NCOIC (non-commissioned officer in charge).

The Mission Recording System (MRS) and post-mission analysis

The MRS had two "decks" for recording mission results. The SR-71 had over 800 sensor monitoring points that, in turn, were connected to "wafers" on the MRS. There were multiple wafers on one spindle, and the spindle revolved every three seconds. One of the decks recorded all analog data, while the other contained a combination of analog and digital data.

When the aircraft landed, maintenance personnel took the MRS tapes to Beale's SAGE Building for formatting and analysis by the Ninth Reconnaissance Technical Squadron (9RTS). The SAGE Building contained the 9SRW Commander's offices, DO, IN, and 9RTS. It was a large concrete-block building with no windows. The building had large diesel generators for backup power should the commercial power fail. There was originally an old vacuum tube computer in the building and if the air conditioning units failed, everyone had ninety seconds in which to evacuate the building to avoid a massive explosion of the vacuum tubes; hence, the backup power.

The MRS tapes were old 3-inch telemetry tapes. They contained the sensors' voltage readings, representing sensor voltage levels from zero to two volts. The data was reformatted to digital 7-track tape by 9RTS technicians, reducing the analog voltage levels to a numeric value of 0-255 for analysis against the sensors' calibration file. The digital data on the tape was processed by additional software that produced printouts for each sensor system and allowed subsequent analysis.

The final version of the MRS was all digital. No A to D conversion was required.

Mission-planning

The pre-mission staff ran the mission-planning software and "route search" software, using a library of "canned" pre-designated routes and adding "incidental targets of opportunity" to the targets already planned. The Control Data 3200 was operated by 9RTS and had punch-card inputs for the mission-planning software. Upgrades to the 3200 software were done by the Air Force and typically involved changes to accommodate the installation of new, improved sensors. The initial SR-71 had an SLR (side-looking radar) that was later replaced with high-resolution radar, but both systems had "stupendous" results.

The "Weight-Balance" Program

The Autoflight Division also ran software to calculate the optimum fuel load. Each SR-71 had slightly different sensor configurations and a log was kept, by aircraft, of the serial numbers of each sensor. Knowing the serial numbers of the on-board sensors allowed the Air Force staff to accurately compute the weight of the sensor load. This, along with the engine-start time of the mission, the take-off time, and other aircraft configuration details, allowed the software to compute the accurate fuel load for each SR-71 tank for optimum balance.

Tony's team and some of their accomplishments

Tony worked for Lieutenant Greg Faber and Greg's boss, Lieutenant Colonel Russ Hixon, the "smartest man" Tony had ever known. Russ had a master's degree in theoretical mathematics and was a software wizard on the CDC 3200. Among Russ' many achievements was a 3200 assembler-level program that would predict the exact coordinates of every photograph taken on the mission. One obvious use was to compare photographs of a specific region over a period of time. Originally, Russ Hixon was a navigator, but never an RSO on the SR-71.

Tony Shelton worked on the team that was involved in the conversion from the 24-bit word CDC 3200 to the 32-bit word SEL 32/55. The SEL computer was chosen as part of the normal AF bidding process.

A side note: The CDC 3200 had only 32K of 24-bit memory. 8K were taken up by the operating system, so that any software only had about 24K in which to operate. Sometimes, when the software was complex, it had to be split up into pieces called "overlays" and brought into memory from magnetic tape, as needed.

Greg Faber and Tony wrote a FORTRAN program to plot traces on a newly acquired flatbed plotter that gave a good "early view" of the mission's results.

Of particular interest to the SR-71 program management was another piece of software they developed to report performance of the inlets on the two jet engines.

All the software developed by the team on which Tony served was based on interviews with personnel in the maintenance areas. Nothing was handed down from "an ivory tower."

Parting thoughts from Tony

The job he had with Greg and Russ was "at least tied with" the best job he ever had. He "got to DO (learn) his job every day, rather than "train for it.""

Tony left the Air Force as a Senior Master Sergeant in 1985 and stayed in IT. He went to work for an insurance company in Nashville Tennessee, later moving to Jacksonville, Florida where he is still employed. He's "not sure he'll ever retire."

The software report card

The CDC 3200 software got an A-plus. Everyone on the project took pride in his or her work, where "perfection became the standard."

A special story from an SR-71 maintenance expert

Reprint from *Air & Space*...

AIR&SPACE INTERVIEW

Don Campbell

An expert on maintenance of the Lockheed SR-71 Mach 3 reconnaissance aircraft, Campbell was the SR-71 Superintendent at both Kadena Air Base in Okinawa and later at Beale Air Force Base in California in the 1970s.

Selected as the first crew chief assigned to the SR-71 program at Beale Air Force Base, California, in 1964, Don Campbell literally wrote the book on SR-71 maintenance: He created the source document used for all SR-71 inspection and replacement criteria for the life of the program.

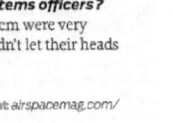

Do you remember the first time you saw an SR-71?
When I was selected for the program in 1964, I was working on the B-58 program, and the first four pilots selected for the SR-71 were B-58 pilots, and they're the ones that got me to go. They showed me a black-and-white picture of an airplane, and that's the first time I saw what I call a Blackbird: It happened to be a YF-12. They asked me if I'd like to go work on an exciting program, and they showed me that picture, and I said sure. My first actual sight of an SR-71 didn't come until early 1965, when I was sent to Palmdale, California, to follow the production.

Was what you saw then a fully assembled Blackbird?
Yes. I got to see some more later. I spent a total of about eight months between Palmdale and Burbank [where Lockheed's Advanced Development Projects division, also known as the Skunk Works, was located], so I got to meet with all the engineers and meet Kelly Johnson [then vice president of Advanced Development Projects].

What was it like meeting Kelly Johnson?
Very humbling. He was a gentleman in every sense of the word, and treated us all with great respect. I'm still in awe of him today. We need another 25 or 30 just like him in this country. I was 25 years old at the time, I think, and I got to be at several meetings with him. It was tremendous.

Some of these SR-71 flights were over hostile territory; because of that, did you and your team feel a lot of pressure to provide the flight crews with a sound aircraft?
Absolutely. The SR-71 is the only aircraft that I've ever been involved with in my 51-year stint in aerospace where the flight crew never looked at the aircraft before a mission. And the U-2 guys, they don't fly at Mach 3, and they can lose an engine on a U-2 and still fly for 45 minutes and land. It's a very safe aircraft—it's a glider. The SR-71 is quite different: When you're flying at 3,200 feet per second, things happen so fast. So because these guys—the pilots—didn't come out and look at the aircraft, my policy as a maintenance superintendent to all of the maintenance guys that worked for me—it was very simple. If I ever catch you doing anything on this aircraft without a checklist or a tech order, you're fired. And the flight crews knew that: When they walked out in the hangar, the crew chief would turn around and give them the thumbs-up, saying the aircraft's ready to go.

Did you ever have to fire anyone?
Yes.

What was your opinion of the SR-71 pilots and reconnaissance systems officers?
Ninety-eight percent of them were very likable individuals that didn't let their heads get too big.

Read the entire interview at airspacemag.com/campbell

Interviews with two of the flight crews

The entire list of SR-71 crewmembers is printed in Appendix A. from **http://www.sr71.us/Supp_BBook.htm**, the description for which reads:

> "The following data lists all personnel who qualified to fly the Blackbirds. The listing is from the first person to fly the aircraft to the last person qualified. The listing includes all RSOs and VIPs that flew in the aircraft. This original data was derived from Buddy Brown's Book and has been expanded upon by David Allison, webmaster at Habu.Org. He has diligently researched and verified those entries contained below. There is data missing but to our knowledge that data is currently not available from any source. As time goes on and data becomes available, this web page will be updated. You can search this page for a name by going to: Edit/Find in Page and entering the name of the crewmember. This page should verify any rumors as to 'The Politician who said he flew a Blackbird.'
>
> The Data Base is Searchable in its entirety
>
> Special thanks to David Allison for making this possible."

The SR-71 normally had a crew of two: A pilot and an RSO (Reconnaissance Systems Navigator). Appendix A lists a total of 155 with title "pilot" and 118 with title "RSO."

The crew of two normally trained together and stayed together for a number of missions and a number of years. Two crews are profiled in this section:
1. BC Thomas (Pilot), Jay Reid (RSO)
2. Gil Bertelson (Pilot), Frank Stampf (RSO)

You will enjoy their stories!

Lieutenant Colonel Bredette (BC) Thomas, SR-71 Pilot,

A quote from BC regarding the AIL software:

"I certainly was an "end user" of the SR-71 mission-planning process, which did, indeed, include the mission-planning software outputs. Our mission-planning for each flight was primarily checking the information presented to us by the computer printouts, coordinating this information with weather, the maps, course (and proximity of threats), as well as fuel expended and fuel remaining. In my experience, I have never known of an error in the mission-planning software, and if you were the lead engineer for that program, I say, 'Thanks for a job well done!!'"

BC commented further that although the SR-71 could be flown manually, the autopilot certainly helped. The cockpit had no onboard computer and was a collection of dials and knobs that required constant monitoring. Being on autopilot allowed BC and other pilots to monitor all of the controls. It was also an Air Force requirement that, in "take areas" (target areas), the SR-71 be in autopilot mode.

SR-71 cockpit

A thumbnail of BC's Air Force career

BC Thomas went to Beale Air Force Base in 1976 after attending Test Pilot School for three-and-one-half years. He had six months of training in a simulator, one hundred hours of SR-71 flight experience and then, after about one year, was ready to embark on his first operational missions. Although he and all the SR-71 pilots were based at Beale, he was assigned missions out of Beale, out of Okinawa, and out of England. The three operational bases, according to BC, conveniently divided the world into three segments. He also ran some flights out of Palmdale, California, where he continued to serve as a test pilot and, additionally, did a number of air shows. He retired from the Air Force in 1987.

Interesting experience #1, in BC's own words and with his photographs

(Note: the following story details what would seem as a true emergency but, due to BC's extensive training, it was a set of circumstances for which he was well-prepared).

That Time an SR-71 Made an Emergency Landing in Norway

On August 13, 1981, Jay Reid and I made the first SR-71 landing in Continental Europe. It was virtually unannounced—and not particularly welcomed.

THE MISSION

Before establishing a continuous SR-71 presence in Europe in 1982 (at RAF Mildenhall, England), the 9th Strategic Reconnaissance Wing, the responsible headquarters for all SR-71 operational flights, was sometimes tasked to fly a particularly important higher-headquarters mission from Beale AFB in California to the Soviet Union and back.

The purpose was to photograph (with either film or radar), and collect electronic data in and around the Soviet Naval facility at Murmansk, located on the Kola Peninsula in the Barents Sea above the Arctic Circle, in the extreme northwest portion of the Soviet Union, north of Norway and east of Finland.

We required information about their air-defense electronic warfare capabilities and, specifically, their anti-aircraft surface-to-air (SAM) missile systems. Murmansk was a strategic nuclear submarine base and maintenance facility, and since knowing the disposition of all nuclear threats was vital for the security of the United States, Murmansk was one of our most significant reconnaissance objectives.

My Reconnaissance Systems Officer (RSO), Jay Reid, and I were assigned this mission scheduled for Wednesday, August 12, 1981 in SR-71 aircraft #964.

We reported for the flight at 7:00 p.m., checked weather and last-minute changes to the mission requirements, updated the intelligence briefing, met with the mobile crew and SR-71 crew chief, ate a meal of steak and eggs, donned our space suits, conducted an aircraft preflight, and were ready for a 9:47 p.m. takeoff.

The flight profile called for a 45,000-pound takeoff fuel load, a rendezvous with two tankers over Idaho to fill our tanks, acceleration to Mach 3+ (2,000 mph) while cruise-climbing to 80,000 feet, descending to 26,000 feet to meet more tankers over Goose Bay, Labrador, and crossing the Atlantic Ocean just south of Greenland and Iceland at Mach 3 to a third refueling over the North Sea, northeast of Scotland.

The next acceleration was to Mach 3.25 to penetrate the Soviet air defense region around Murmansk. Passing west of Norway, then north of Finland, we headed toward the Soviet coast by flying inbound from a point in the Barents Sea and approaching the landmass at Murmansk directly at 90 degrees head-on.

There is no doubt that our presence was heralded to Soviet personnel on the ground by the sweet (and rather loud) sound of freedom: the distinctive double-sonic boom of the 2,200 mile-per-hour SR-71.

That knowledge would always give us a certain inward sense of pride; not in ourselves altogether, but pride in the aircraft, the engineers and designers who built her, the maintenance people who made her safe to fly, and the faithful tanker crews who were always there when we needed fuel. All of these support personnel were vital for any successful mission.

The Soviet Union claimed their sovereign territory extended 100 nautical miles from their landmass. The international norm is 12 miles. Heading inbound, we turned so that we flew within 12.5 miles of the Soviet coast in a 30-degree right-banked turn while obtaining radar imagery (or photographs) and recording Soviet electronic countermeasures.

This somewhat provocative technique was trolling: we stimulated the Soviet defenses, causing their radios and radar to bristle with electronic information, thus impelling them to reveal telltale electromagnetic signatures indicating the type of equipment, modes of operation, and limits. There were sometimes other American assets offshore, but within radio range, which also collected electronic transmissions.

After departing the "take area," the plan was to refuel again over the North Sea, accelerate to Mach 3, refuel for a fifth time over Goose Bay, then fly Mach 3 thereafter to Beale AFB. We planned to land at 8:15 am on August 13 after 10.5 hours of very intense flying.

We were operating a long distance from home base, the mission was deemed vital, and no crewmember ever wanted to abort a flight. Any mission deviation could lead easily to an international incident, would be scrutinized by the highest National Command Authority, and possibly result in our names appearing in every major newspaper in the world.

We certainly felt this responsibility, but had the utmost faith in our preparation, training, and the talent and dedication of all our support branches.

THE EMERGENCY

After transiting Murmansk and while hooked up with the tankers during the fourth refueling, I saw a flicker of the Master Warning light.

Jay Reid announced the light at the same time that I saw it in my peripheral vision. I disconnected from the tanker's boom and maneuvered back to the pre-contact position. Our indication was the illumination of the left-engine oil supply low-quantity red warning light. From our training and experience with the aircraft emergency checklists, we knew immediately that this required that we "land as soon as possible."

The mandate to land immediately was borne out of long experience with malfunctions: the type that would crash the airplane if another single-system failed.

The natural tendency for military aircrews is to complete the mission if humanly possible. To counter this inclination, the Wing Commander had designated certain emergencies sufficiently critical to require immediate landing. This was one of those emergencies.

Our mission-planning and alternate airfield study had prepared us to know quickly where to fly if an emergency required immediate landing. Some military intelligence might limit our choice of a suitable landing field, depending on the political situation, and of course weather was always a major consideration.

Jay and I quickly agreed that recovery at the joint civilian/military air base at Bodø (pronounced: "Buddha"), located on the Norwegian coast a few miles above the Arctic Circle, was best. We were about 60 miles away.

THE ARRIVAL

We wasted no time notifying the tanker crews of our decision to land. They set up a holding pattern in international airspace to assist in refueling, or to be our radio-relay if necessary. Their standing orders were to await our call to release them, presumably after our safe landing.

While Jay was transmitting our mandatory abort reports via high frequency, long-range radio, I contacted Norwegian Approach Control. Our orders were not to broadcast that we were flying an SR-71, but rather give the general type as "U.S. Air Force Tactical," which really meant nothing to a controller concerned about arranging for emergency equipment and notifying proper authorities of our emergency condition.

My call sign was "Belmont 86" and my transmissions to approach control were something like this:

"Bodø Approach, Belmont 86, six-zero miles west, declaring an in-flight emergency, request straight-in approach to land on runway zero seven."

"Belmont 86, say aircraft type, nature of emergency, souls-on-board, and fuel remaining."

"Bodø Approach, Belmont 86, US Air Force Tactical, engine problem, 2 souls, zero plus four-five fuel-on-board."

"Belmont 86, say again aircraft type."

"Bodø Approach, Belmont 86, US Air Force Tactical."

"Belmont 86, I do not understand your aircraft type."

"This is Belmont 86, we will land in approximately 10 minutes and I will deploy 3 drag parachutes: 2 small and one 40-foot chute. I will jettison all three on the runway. I don't have time now to talk."

We were cleared to land with nothing more said. We touched down at Bodø on August 13 at 1:12 p.m. Norwegian time after a total flight of 6.4 hours.

When we were on short-final approach, tower controllers had no problem identifying our aircraft type, and immediately sent out notifications that, we learned later, included the local press.

What could we expect? What kind of reception would the Norwegians give us, and would the airplane be secure, especially considering that we had on-board highly classified images and electronic recordings from Murmansk?

THE RECEPTION

I asked for and was given taxi directions to the Norwegian military ramp where I could see some friendly looking F-104s, the front-line NATO fighter aircraft. After I shut down the engines and we opened our canopies, the first person to greet us was a Norwegian military pilot who said: "Welcome, do you know Bill Groninger?"

Bill Groninger was a fellow SR-71 pilot who was a USAF instructor pilot before he was chosen for the SR-71 program. The Norwegian pilot was his student. We learned that many Norwegian fighter pilots received their initial training in the USA. We definitely were among friends!

We were quickly introduced to General Olav Aamoth, the Wing Commander at Bodø. He asked for any special requirements and I requested full-time guards be placed on the aircraft with only myself and Jay Reid authorized to admit personnel to the plane. He agreed and assured us that the aircraft would be guarded continuously. I then asked for a secure phone to call the Command Post at Beale AFB.

General Aamoth drove us to an underground labyrinth of tunnels carved into the side of a mountain containing maintenance shops and aircraft. He directed me to a telephone within his Command Center.

I called the 9SRW Commander, told him the time of the abort and that operations were normal up to the emergency, thus relaying that we had the reconnaissance data on board. This information was needed to arrange for the proper retrieval of the mission materials.

We briefly discussed the nature of the emergency, and I assured him that the aircraft was safe. He said that we were to stay there until the aircraft was repaired, which would probably take three days. The 9SRW had already started a recall for the support personnel necessary for the recovery of the SR-71.

General Aamoth seemed unusually concerned, as he insisted that a Norwegian officer physically be present with us until we left. He introduced us to an F-104 pilot, 1 Lt Roar Strand, who would help us with anything we might require. He was a very pleasant guy, and I kind of felt sorry for him because I was sure he had better things to do than to "mind" us for 3-4 days.

Roar Strand was very accommodating. After all the immediate activities associated with bedding down the aircraft were accomplished, he took Jay and me to his apartment, where we met his beautiful girlfriend, then to a restaurant in downtown Bodø where we spent a pleasant evening.

Roar accompanied us to the transient quarters to sleep in the same room with Jay and me. Early the next morning, he went with us to a military dining hall for breakfast. It was there that I got my first real surprise of the trip: the only food presented was about four types of fish soup, all of which, to this unaccustomed American, smelled terrible, especially so early in the morning. I asked if any cold cereal was available and, thankfully, there was – Corn Flakes. Espresso coffee completed the fare.

General Aamoth joined us and with him was the American Air attaché to Norway who had just flown in from Oslo specifically to see us. I knew then that this visit was attracting more attention than any of us needed or wanted. He asked about the emergency, the condition of the airplane, and wanted to know if anything unusual or unpleasant had happened to us.

I assured him that we were fine and that recoveries such as this one were well planned and the maintenance team would probably be here that day or the next. We did not anticipate any logistical problem as this type of activity was performed often, usually without help from other local organizations, and the Norwegians were taking extra care to assure that we were not contacted or hindered.

RECOVERY OPERATIONS

The rest of the day was devoted to making preparations for the maintenance recovery team from Beale AFB. General Aamoth, to my surprise, stated specifically that he did not want any talk or outward expression of anything clandestine (not that we would anyway).

More than once he made the explicit statement that when we landed, he considered us to be a NATO-allied aircraft in distress and that military courtesy and professional consideration meant that we were to be afforded the support necessary to see us on our way and that he did not want any mention of reconnaissance activity or publicity.

It was way too late for the admonition concerning publicity, however.

Virtually every newspaper in Norway had news of a "Spy Plane" landing at Bodø splashed across its front page. Newspapers in the USA also had the story. The reports generally said that it was the highest-flying and fastest airplane in the world and opined on what type of "spying" we were doing.

Jay and I did not like the publicity but were relieved that no newspapers published our pictures or names. None of us flying reconnaissance missions wanted to be publicly identified while we were engaged in these activities.

Our missions were highly classified, and no good could ever come from talking to reporters about anything associated with our reconnaissance flights.

The next day, August 15, a KC-135Q from Beale AFB arrived carrying our special fuel for the SR-71, the deployment commander, Lt Col Randy Hertzog, who was also the Commander of the 1st Strategic Reconnaissance Squadron (1SRS), and the maintenance crew, plus civilian technical representatives from Lockheed.

The recovery plan written by the 9SRW stated that in foreign stations, the recovery team would all wear civilian clothes so as not to disclose who was military. This did not sit well with General Aamoth! He told me to tell Col. Hertzog to have all military personnel wear military uniforms.

I did not know why he would issue such an order, but I complied. The military team dutifully returned to the KC-135 and changed into uniforms.

Over the next two days, the maintenance team ran into several problems repairing the aircraft. We had parts flown in from various sources. A spare "engine start cart," a machine containing two Buick Wildcat engines in tandem and specifically designed to mechanically rotate the SR-71 engines for starting, was flown in by C-130. On the third day, our aircraft was repaired and made ready for takeoff.

Since preparing the SR-71 for a supersonic (Mach 3) flight would have taken more equipment and time, we elected to fly subsonic directly to RAF Mildenhall, England, from which the SR-71 had conducted operations previously. We would fly in formation with the KC-135Q, which had landed at Bodø for our support two days earlier.

DEPARTURE

The maintenance crew performed outstandingly and we were ready to depart on Sunday, August 16. Preflight preparations for Jay and me were smooth, except our first engine-start attempt failed. The start cart could not, observing normal rpm and torque limits, achieve the required aircraft engine speed for a safe start.

"On the second attempt, the crew chief, Clarence "Skip" Hosler, valiantly ignored the cart rpm/torque limits and informed me on inter-phone that this would be the last attempt. He later told me that the connecting rod glowed red-hot just before the aircraft engine reached the start rpm.

We took off at 1:42 p.m. It was the roughest takeoff I ever made in an SR-71. The runway was smooth to sight, but owing to the very long, slender and flexible fore-body of the SR-71 (where the nose wheel is located), and the distance from the nose wheels to the main landing gear (42 feet), the slight undulating wavelike runway set up a vertical motion in the cockpit which amplified as we accelerated for takeoff.

It was getting so bad by the time we achieved rotation speed (180 knots) that I was very concerned that something might break before liftoff (210 knots). The acceleration of the SR-71 was quick and I was able to raise the nose before the up-and-down oscillation became traumatic. With the gear up, everything returned to normal. We rendezvoused with our tanker for a planned refueling to augment our takeoff fuel, and landed at 2:53 p.m. local time in England, for a flight of 2.2 hours.

With a million members of the Polish Solidarity movement having gone on strike on 7 August and mounting tension between Communist state officials and the rest of the Polish population, Strategic Air Command (SAC) directed that, upon arrival, we would set up a deployment base at RAF Mildenhall and conduct several additional operational missions before returning home.

Jerry Glasser and Mac Hornbaker were sent from Beale AFB to meet us when we arrived from Norway, and they flew an operational mission a few days later. We were ordered to stay at RAF Mildenhall until relieved, so Jay and I also flew another operational mission after Jerry and Mac. When Rich Young and Ed Bethart arrived as our replacements, we departed for Beale AFB on September 2.

What started out for us as a 10.5-hour, round-robin mission, California to the Soviet Union and back, turned into a 21-day marathon of SR-71 missions, involving six SR-71 air crew members, a host of maintenance and support personnel, and a true adventure for us all.

Along the way, we made some very special friends.

AFTERWARD

After I retired from the US Air Force and our SR-71 reconnaissance activities were no longer classified, I contacted General Aamoth. He had also retired from the military after being Chief of Staff for the Norwegian Air Force from 1985 through 1991.

I asked him about his reluctance to have our military personnel wear civilian clothes and why he admonished us not to do or say anything that would identify our mission as being reconnaissance. He told me a very interesting story.

On May 1, 1960, General Aamoth was a fighter pilot (then a Lieutenant) in the Norwegian Air Force, stationed at Bodø. That was the day that Francis Gary Powers was shot down flying a CIA U-2 aircraft over the Soviet Union. Although Powers had taken off from Pakistan, his intended landing base was Bodø, the same Norwegian base at which I landed.

Furthermore, the then-commander of Bodø knew about the U-2 and that it was to land at his home base; however, neither the Norwegian Prime Minister nor the Chief of the Norwegian Air Force knew anything about it. To make matters much worse, the Soviet Premier, Nikita Khrushchev, threatened to "nuke Bodø" for "cooperating" with the United States in the 'U-2 Affair.' After an investigation, the Military Commander was summarily fired.

General Aamoth saw a very stark parallel to that story when we landed. He told me that he was sitting in his office when we were on final approach. The airfield tower controller telephoned and told him to look out his window to see what aircraft was landing at his field asking to be parked on his military airport apron.

I now understand General Aamoth's concern!

As of 2016, Roar Strand is still an active pilot, flying worldwide for Scandinavian Airlines (SAS). He has homes in Norway and France and is married to Maryann, the girlfriend whom I met at his apartment in 1981. They are the proud parents of three and are also grandparents.

As for SR-71 aircraft #964, a veritable workhorse for the SR-71 program, it is displayed at the Strategic Air and Space Museum (formerly the SAC Museum) near Omaha, Nebraska.

John Morgan (RSO) and I had an SR-71 emergency on April 5, 1984 in aircraft #974 and landed again at Bodø. General Aamoth was there to meet me a second time. He was then a Major General and Commander of Tactical Air Forces North. I told him that I really didn't mean to do it a second time! There was minimum publicity and we departed within two days. By then, we were operating out of RAF Mildenhall and the logistical support was much closer and quicker than in 1981.

The SR-71 (#974) in which I landed at Bodø the second time was the first SR-71 to fly combat missions in the Vietnam War. It was also the last Blackbird to crash. While flying at Mach 3 over the Philippine Sea on April 21, 1989, her left engine seized, causing shrapnel to explode into critical hydraulic lines and subsequently rendering the flight controls completely inoperative. The crew, Dan House (pilot) and Blair Bozek (RSO), ejected safely and were recovered by Filipino fishermen.

By May 10, 1989, US Navy and Air Force personnel had recovered the wreckage from 200 feet below the ocean surface and returned her to Okinawa for disposal."

Interesting Experience #2:
Four Concurrent Emergencies on Landing

Because air refueling was both difficult and dangerous, BC welcomed an assignment that did not require refueling. It should have been easy – A 45-minute flight over North Korea then back to Okinawa. It was his first tour in Okinawa and he had been an SR-71 pilot for about one year.

Russian and Chinese trawlers patrolled the seas near the base at Okinawa and carefully monitored all U.S. flights. Because of the nature of this particular mission, it was decided that the flight occur with a higher than normal load of 65,000 pounds of fuel, and under radio silence.

BC encountered an engine inlet problem resulting in a series of uncontrollable "unstarts" and had to abort the mission, returning to Okinawa after just twenty minutes and still with 15,000 pounds of fuel. This was just the beginning of BC's problems. Because he returned to Okinawa much earlier than expected, the crew responsible for maintaining the hangar did not have time to clean up the fuel slick on the hangar's floor.

An SR-71 always lands with an eleven-degree nose-high attitude, deploys a drag parachute, and then eases the nose down for a smooth landing. On this particular landing, however, the nose went down, and the plane started to vacillate left and right and knock out the right generator. Further compounding these difficulties, and because BC's arrival was way ahead of schedule, the crews did not have time to clear away the "arresting cable" on the runway. BC tried to slow down but could not; he crossed over the arresting cable and blew out two of the right tires.

The normal recovery procedure was to go into the hangar with the engines running and then cut them off. He got into the hangar at three MPH and, because of the oil slick, could not stop the plane. He tried the brakes, to no avail. BC had two seconds to maneuver the airplane and decide whether or not to cut off the engines. He forced a strong right turn and the aircraft came to a halt, unharmed.

Intentionally Not Learning the SR-71 Sensors' Capabilities

BC chose NOT to know these specifics of the sensors' capabilities because then, had he been captured, he could not be coerced into revealing them.

Final comments on the SR-71

"All the money went into the sensor systems. They did not make the SR-71 easy to fly!"

BC's life after the Air Force

BC went to work for Northrup after leaving the Air Force in 1987. He was hired by Northrup's chief B-2 test pilot, Bruce Hines, with whom he had served as co-pilot on a C-130 in Vietnam. The B-2 project was highly classified, and Bruce was unable to initially tell BC the nature of his assignment. The B-2 was not yet ready for flying, so BC was assigned cubicle duties as a Flight Test Engineer, something that got tiresome after six months. After about one year, BC left Northrop for a job as Test Pilot for United Airlines, where he stayed for fifteen years flying "broken" planes in for repair, testing all of Boeing's new aircraft and, lastly, test-flying "rehabilitated" aircraft. Every five years, a plane has an HMV (Heavy Maintenance Visit) in which a plane has to be disassembled and reassembled with the latest maintenance equipment (FAA rules). BC's job was to test those planes after the reassembly process and do a Heavy Maintenance Test flight.

Photos from BC's Private Collection

BC's last official photograph

More from the collection

SR-71 takeoff

SR-71 landing

SR-71 climbing, full afterburner

Lieutenant Colonel Jay Reid – SR-71 Reconnaissance Systems Officer (RSO)

Jay Reid was interviewed for this book and provided the following recollections of his time at Beale Air Force Base, California, some of his SR-71 experiences, and life after flying.

Flying Career – From Beale to Beale to Beale

Jay was assigned to Beale AFB from 1971-1982, first in B-52s sitting "nuclear alert" and then as a navigator and navigator/bombardier during the Vietnam era where he completed over 100 bombing missions – including a near miss from a Russian MIG fighter missile attack. In 1975, following the Vietnam War, Jay took a career-broadening assignment with the 9th Strategic Reconnaissance Wing (SRW) at Beale, first in the SR-71 Avionics Maintenance Squadron commanded by Colonel Jon Kraus, then as the 9SRW Squadron Section Commander, reporting to General John Storrie, Wing Commander, where he served as Administrative Commander of the 9SRW enlisted personnel. During this assignment, Jay applied for the SR-71 RSO flying position. Upon selection, Jay was crewed with BC Thomas (pilot) and began the intensive yearlong training program of academics and simulator immersion to achieve the necessary skills and crew coordination to qualify for their first SR-71 flight. Jay then flew with BC Thomas for over three years, out of all three operational bases: Beale Air Force Base (California), RAF Mildenhall (England), and Kadena Air Base (Okinawa). Jay then flew for another 18 months with Tom Allison (SR-71 pilot).

"You've never been lost until you've been lost at Mach 3"

Although he may never have been lost, it's a great quote – flying a mile every two seconds or three times the speed of sound requires constant vigil. According to Jay, the mission-planning and navigation software specifically programmed for each SR-71 flight performed flawlessly. However, in addition to normal flight crew duties, the RSO was responsible for monitoring all sensor and navigation activity to assure that the aircraft precisely followed its planned flight route and remained on track (especially in the take areas) where the sensors performed their critical collection.

Lightning Strike

SR-71 crews flying 10-12-hour operational reconnaissance missions from Beale to Europe and back routinely did as many as five refuelings, taking on fuel from two KC-135Q model tankers. The crew would descend from above 80,000 feet to rendezvous with the tankers at around 26,000 feet – sometimes in daylight and clear weather and sometimes at night in less than ideal weather. With no weather radar to identify thunderstorms at night, the crew relied on macro/regional forecasts and tried to avoid any thunderstorms visually illuminated by lightning. Jay recalled one particular descent into lightning and thunderstorms where, upon landing, the crew was advised by the recovery maintenance personnel that the aircraft had been hit by lightning on one of the vertical stabilizer tails.

The P-3 Sonar Buoy Incident

The Thomas-Reid crew flew an operational mission out of Okinawa and was returning to base as a Navy P-3 Orion landed just ahead of them and inadvertently dropped a sonar buoy on the runway. The SR-71 nose wheel hit the buoy on touchdown and both Jay and BC experienced another never-to-be-forgotten moment as the aircraft pitched violently over the buoy and BC expertly controlled the aircraft to complete the landing sequence. The SR-71 buddy crew retrieved the sonar buoy, and the SR-71 crewmembers and Detachment Commander (Col Rafael Samay) subsequently presented the sonar buoy on a plaque to the Navy flyers and their P-3 Squadron Commander at a slightly embarrassing ceremony. The ceremony was followed by a brief happy hour at the P-3 Squadron where additional war stories were exchanged, and BC and Jay were invited to go along on a future P-3 submarine tracking mission.

The Kadena Hangar Slide

The Thomas-Reid crew experienced numerous airborne emergency situations such as the well-documented SR-71 abort into Bodo, Norway. However, BC and Jay also experienced some scary situations on the ground. They completed one operational mission and were taxiing up to the Kadena hangar that is open on both ends. The SR-71 taxied into the hangar where the crew chief routinely waits at the other end of the hangar. As BC taxied the SR-71 into the hangar, the floor was covered with JP7 fuel from flight prep and prestart for this mission. There is no fuel bladder in the SR-71; the skin of the aircraft is the skin of the fuel tank. JP7 leaks out of the aircraft throughout ground operations and until the aircraft is airborne, where it heats up and the elastic polymers expand and seal off the leaking JP7. An SR-71 can leak as much as 1000 pounds of fuel between preflight and takeoff.

As the SR-71 slowly taxied into the hangar and approached the parking spot, the crew chief signaled BC to stop. BC applied the brakes as usual, but the SR-71 began to slide forward in slow motion on the JP7 fuel residue. The Physiological Support Van, which transports the crew back to the facility where they get out of their pressure suits, is normally parked in front of the aircraft at this point (between the nose of the aircraft and the blast fence just outside the hangar). Fortunately, it was not parked in its usual position or it would have been hit as the aircraft continued to slide on through the hangar.

The SR-71 buddy crew (Rich Graham and Don Emmons) and other ground personnel immediately placed chalks in front of the wheels and everyone on the ground grabbed some part of the aircraft, attempting to help stop it.

As Jay stated, "The problem of not hitting stationary objects while sliding through the hangar was complicated by having to make about a 30-degree right turn to avoid the blast fence before the tires were on firm, dry ground and we could stop. A turn too early and the right wing would contact the hangar wall; too late and the steering might not work, causing us to crash into the blast fence (it's for things like this that we were paid the extra bucks). After several moments of stark terror, the aircraft finally stopped sliding as the landing gear reached dry concrete. As they say, it's not over until it's over – in this case the flight was not over until the aircraft had completely stopped halfway outside the hangar. After this incident, the hangar floor was thoroughly mopped up prior to parking to prevent any future incidents."

The Air Force after Beale

Following his SR-71 tour of duty, Jay was assigned to the Joint Reconnaissance Center (JRC) in the Joint Chiefs of Staff (JCS) at The Pentagon, Washington, D.C. Jay worked 12-hour shifts tracking worldwide reconnaissance activities and advising on the SR-71's capabilities within the National Military Command Center. Jay coordinated the Plans Branch of the JRC, where he planned and participated in war game "exercises" with SAC Headquarters staff. He served at The Pentagon from 1982 to 1984 before being reassigned to the Logistics Command at McClellan Air Force Base, California.

At McClellan AFB, Jay supervised a Contract Management Division within a large data center, doing Logistics Command payroll processing for facilities up and down the west coast. Jay advanced to Acting Commander for the Organization until his retirement as a Lieutenant Colonel in 1993, culminating a successful 24-year career with the Air Force.

After Air Force Retirement

Most importantly, Jay met his future wife, Pamela, during his last assignment and they married in 1992, when Jay was 47. He obtained his real estate license and worked as a loan officer while managing an international portfolio of residential and commercial properties. After that, Jay accepted a position as Executive Officer for the Western Electrical Contractors Association, coordinating activities for 200 contractor members and a state-approved electrician apprentice training program with 12 regional training sites. In 2002, he was recruited to work for the California Department of Education where he served as an Education Programs Consultant in Data Management until retiring again in 2016 at the age of 71.

Jay has three stepsons, Jeff, Greg, and Tim, and seven grandchildren. Pamela and Jay live in Rocklin, California. They serve at their church "Society (Christian) Church" on Sundays and spend time at their Lake Tahoe cabin. According to Jay, looking back on his life and flying experiences, he "feels like a cat with nine lives that has used up about eight of them."

I would say he used them well!

Colonel Gil Bertelson – Surviving an engine fire and the experience of three sunrises on one particular day

Gil Bertelson, like all the other SR-71 pilots, had his share of good stories. He was an F-111 pilot flying out of RAF Lankeheath in the United Kingdom just prior to his selection to the SR-71 program. In addition to SR-71 missions being flown out of Kadena Air Base on Okinawa, Japan, some missions were also flown out of RAF Mildenhall in the United Kingdom. As the crow flies, RAF Lakenheath and RAF Mildenhall were only about three miles from each other.

During his first deployment to fly the SR-71 out of RAF Mildenhall, he had a chance to have dinner with some of his F-111 squadron mates who were still assigned to RAF Lakenheath. During dinner, his friends began to good-naturedly badger him about how high and how fast the SR-71 could really fly. At that time, the actual figures were still classified. However, because of validated information obtained on planned record setting flights of the SR-71, there were published numbers stating 85,000 feet and 2,197 MPH. But the friends wanted to know how high it could really fly. They suspected, rightly so, that the record setting numbers were not the ultimate speed and altitude it could obtain.

There is a famous aviation poem called "High Flight", written by a young American who, prior to the United States entering World War II, was flying with the RAF. The poem reads as follows:

> *Oh! I have slipped the surly bonds of Earth*
>
> *And danced the skies on laughter-silvered wings;*
>
> *Sunward I've climbed, and joined the tumbling mirth*
>
> *Of sun-split clouds, — and done a hundred things*
>
> *You have not dreamed of — wheeled and soared and swung*
>
> *High in the sunlit silence. Hov'ring there,*
>
> *I've chased the shouting wind along, and flung*
>
> *My eager craft through footless halls of air… .*
>
> *Up, up the long, delirious burning blue*
>
> *I've topped the wind-swept heights with easy grace*
>
> *Where never lark, or ever eagle flew —*
>
> *And, while with silent, lifting mind I've trod*
>
> *The high untrespassed sanctity of space,*
>
> *Put out my hand, and touched the face of God.*

Gil's response to the queries was, "I can't give you exact numbers, but I can give you something to relate to … you know the part in "High Flight" where it talks about putting out your hand to touch the face of God? Well, when we were "at speed and altitude" in the SR, we had to slow down and descend in order to do that."

The Limiting Factor for the SR-71

Believe it or not, the limiting factor in the speed of the SR-71 was what was known as CIT – Compressor Inlet Temperature. The maximum allowable CIT was 427 degrees Centigrade. Above that temperature, the engines would begin to burn up. On a "standard day" (temperature wise), the outside air temperature at the SR-71's operating altitudes would be minus 56 degrees Centigrade. On one of those "standard temperature" days, the CIT would reach the limiting 427 degrees as the aircraft was accelerated to Mach 3.2. If the outside air temperature was warmer than standard (say minus 52 degrees), the CIT would reach 427 degrees prior to reaching Mach 3.2. Conversely, if the outside air were colder than standard (say minus 60 degrees), the 427-degree CIT would be reached at a higher Mach number. There were occasions, especially when flying out of Okinawa, where the outside air temperature was considerably warmer than standard. If it was warm enough, missions might have to be aborted due to fuel considerations. The airplane was not as fuel efficient in warm air as it was in cold air.

Gil remembers one mission flown over the Barents and Arctic Seas when the outside air temperature was showing minus 90 degrees Centigrade, or 34 degrees colder than standard. Due to that extremely cold air, the airplane reached 90,000 feet and the slowest speed he could maintain was 3.15 Mach. Turn starting points and bank angles had to be manually adjusted to avoid getting too close to the Soviet Union's land mass. When they rendezvoused with the tankers for their next in-flight refueling, they had 12,000 pounds (nearly 2,000 gallons) more fuel on board than the flight plan called for. Those J-58 engines loved cold air.

Celestial navigation during daylight

The CDC 3200 computer (and its follow-on SEL computer) supplied the navigation system with a "nav Mylar tape" of the flight path and target coordinates. This Mylar tape went into the Nortronics Navigation System.

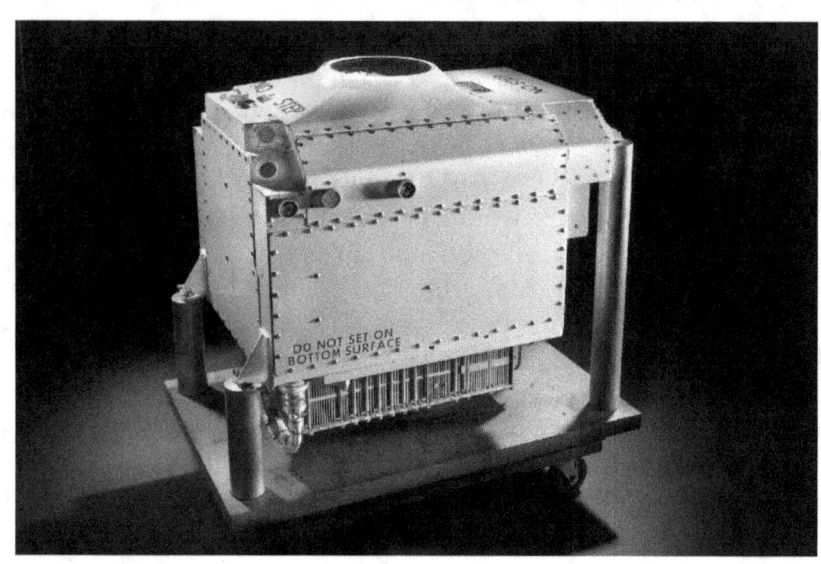

CAPTION:
The Astroinertial Navigation System provided rapid celestial navigation fixes for the SR-71.
TYPE: Artifact
IMAGE DATE: December 17, 2012
Learn more about the Astroinertial Navigation System »
CREDIT: National Air and Space Museum, Smithsonian Institution.
ORIGIN: National Air and Space Museum, Smithsonian Institution.
CREATOR: Eric Long
NASM2013-00438

Yes, Magellan navigated by the stars, but not during daylight and not at 2200 MPH! Rumors had it that the navigation system was so accurate that the SR-71 was never more than 300 feet off its intended/planned flight path.

Interesting Experiences in Flying with Frank Stampf

Gil always flew with Frank Stampf as his RSO (Reconnaissance Systems Officer) and did so for three and one-half years. One of the most memorable experiences, the story of an engine fire, is incorporated in one of Rich Graham's several SR-71 books; this story will not be duplicated here, and we urge readers of this book to check out Rich Graham's books on Amazon.

Gil mentioned one mission that was flown in the late afternoon into evening. There were several in-flight refuelings and several climbs to around 85,000 feet. Due to the route of the mission, on two occasions they actually climbed out of darkness heading to the West and caught up with the sun to witness sunrises in the West. So, on that particular day he got to witness three sunrises – one in the East as the day began and then two in the West as he was outracing the sun.

And after the SR-71...

Gil left the SR-71 program about the same time that his RSO (Reconnaissance Systems Officer) was assigned to attend what was known as a Joint Service School. With Frank's departure to attend the school, there would be some shuffling among the crewmembers in order to fill Frank's void. He was not excited about the prospect of having to fly with someone other than Frank Stampf, plus there was a good probability that he would have to become the primary simulator instructor, something he really didn't want to do.

While looking through an Air Force assignments manual, he happened onto a page that indicated there were staff positions for pilots at CINCPAC (Commander-In-Chief, Pacific Command) in Honolulu, Hawaii. He made a phone call to the number listed in the manual. Initially, there was a cool reception from the officer on the other end of the line. At some point, the officer casually asked what airplane Gil was flying. When the officer was told "the SR-71," things changed in a hurry. The next response from the assignments guy was, "I've got a job for you in February."

Initially, there was some reluctance on the part of the SR-71 program to release Gil for the CINPAC assignment. He was at a point in the SR-71 where he had begun the process to be upgraded to instructor pilot status. After some foot dragging by the SR program to release him for the Hawaii assignment, the personnel center over-rode the foot dragging and executed the orders sending him to CINCPAC. He had mixed emotions about the new assignment. Being a SR-71 pilot was a once-in-a-lifetime opportunity and he had reservations about leaving. However, with this move, he could now spend more time with his family, important for him because, up until then, he was only home fifty percent, or less, of the time. He'd no longer miss out on his four daughters' important life events.

After four-and-one-half years "in paradise," he was transferred to 15th Air Force Headquarters at March Air Force Base in Riverside, California for three years. Following that, he was assigned as the commander of the Air Force ROTC detachment at Brigham Young University (BYU) in Provo, Utah.

Following three years as the ROTC commander, he retired from the Air Force and went to work for the Business School at BYU. He became the director of the Executive MBA Program. The program was designed to allow working professionals the opportunity to obtain an MBA degree while they continued in their current occupations. He spent fourteen years in that assignment and then retired from BYU.

Gil was born in Provo, Utah, just one and one-half miles from where he now lives. He goes fishing in his boat two days a week, plays golf two days a week, and does volunteer work two days a week. He just finished writing his personal history for the National Archives.

A telling quote about the genius of Kelly Johnson: "Kelly's SR-71 was so far ahead of its time, it was retired before time could catch up with it."

Colonel Frank Stampf – SR-71 RSO from 1979 until 1983

"There's nothing normal about flying that airplane."

Before being accepted into the SR-71 program in 1979, Frank accumulated seven years and 1600 hours as a Weapons Systems Officer (WSO) in the RF-4C, the reconnaissance version of the McDonnell Douglas Phantom II fighter aircraft. After acceptance into the SR program, Frank always flew his missions with Gil Bertelson as his pilot. Gil came to the SR-71 crew force from the F-111. Frank mentioned, in fact, that all the SR-71 crews stayed together for the duration of their time in the SR-71 program, with few exceptions. Frank and Gil flew out of all three bases: Beale Air Force Base in California, RAF Mildenhall in England, and Kadena Air Base in Okinawa, Japan. They still see each other socially at every opportunity.

Gil and Frank both entered the SR-71 program as captains. One night during a training flight, while cruising at about 78,000 feet off the coast of California, they were just about to begin their descent to recover back at Beale AFB, when they received a call on the HF (High-Frequency) radio. Rich Graham, then the Squadron Commander, told them that they had both been promoted to major. A pretty cool way to get good news like that.

Frank said that in his experience flying the SR-71, the mission-planning / navigation software performed flawlessly; he never had a need to override the sensor controls and the navigation tape never had a failure (except, of course, in the simulator, where the crews drilled endlessly on how to react to any abnormality, including sensor or navigation system malfunctions).

Frank's most interesting story

The details of his most harrowing experience (an engine fire) are explained in Rich Graham's book, "SR-71 Blackbird: Stories, Tales, and Legends," which we urge you to read.

This story, however, took place flying out of Okinawa. Frank and Gil had just completed their compulsory ten to twelve months of crew training and were slated to fly their first night operational mission over the Korean Demilitarized Zone (DMZ); it was a "bow-tie" (double loop) mission over the Korean peninsula, scheduled to make several passes over the DMZ.

They were on the east side of the peninsula, cruising at Mach 3 on a completely moonless and black night. Gil guided the airplane into a 33-degree bank to the right when they spotted what appeared to be thousands of lights below. This was most unusual because the Korean peninsula was normally pitch black. The navigation systems appeared to be working well, but the lights were a complete mystery and a bit unnerving to a new crew flying very close to some not-so-friendly territory.

They successfully completed the mission, landed the aircraft, got debriefed by the Intel Officers, and found out that what they had seen were huge numbers of small Korean fishing vessels – thousands of sampans with their lanterns lit that gave the appearance of a spread-out city …. something they were NOT supposed to be seeing at that particular moment in time!

Frank was known among his SR-71 squadron mates for the saying: "You haven't been lost till you've been lost at Mach 3!" Additionally, he came up with this corollary to the quote: "….because for every minute that you don't know where you are, you're 35 miles further away from where you think you're supposed to be!"

The 1st Strat Recon Squadron's "recreation room" (read "squadron bar") had a display wall containing the crew pictures of all paired operational SR-71 crews. That room, including the photo wall, was dismantled and later "re-created" in the US Air Force Museum at Wright-Patterson Air Force Base. In all, there were only about 86* operational crews that spanned some twenty-five years of the "Blackbird's" service to the nation.

*As Frank states, "I use the term "about" 86 operational crews because, like me, several crewmembers had to leave the SR-71 crew force sooner than they'd liked due to re-assignments and, in some cases, the remaining "crewmate" would join up with another new partner, thus impacting the ability to say exactly how many operational crewmembers eventually served. This happened only very rarely. Suffice it to say, the entire SR-71 crew force was, in total, a much smaller group than the astronaut corps."

Frank's Air Force career after 1983

After less than four years in the SR-71 program, Frank was selected for the (all armed services) Armed Forces Staff College in 1983. He asked for an operational deferment to delay going to school because, at the time, there were only nine SR-71 crews qualified to fly operational missions, and the cold war tasking for reconnaissance was heating up. Unfortunately, the deferment request was denied and he went to the Virginia-based college, finishing in the summer of 1983.

From there, Frank was assigned to Strategic Air Command Headquarters in Omaha, Nebraska, at the Strategic Reconnaissance Center (SRC). He was promoted to Chief of the SR-71 Operations Branch within the SRC, coordinating worldwide operational tasking for the SR-71. He did this for three years, from 1983 until 1986.

Following his SRC assignment, Frank was selected to be the DO (Director of Operations) of Detachment 4, 9th Strat Recon Wing, at RAF Mildenhall in the UK, which was the SR-71's forward operating location in the European theatre. He served in that capacity for two years (1986-1988) and "enjoyed it thoroughly."

Frank was hopeful that he'd be considered to take over command of Det 4 when the current commander rotated back to the States. However, once again his SR-71 career was interrupted when Frank was selected to "go to school," this time to Air War College (AWC) in Montgomery Alabama for the one-year senior service school program. Rich Graham, the 9th Strat Recon Wing Commander at the time, again asked for a deferment for Frank but, once again, that deferment was denied. At this point in time, the SR-71 program was winding down and Frank's involvement with the program came to an end when he graduated from AWC in 1989, ten years after he had entered the program.

After finishing the year at the Air War College, Frank was assigned to Air Force Space Command at Peterson Air Force Base in Colorado, serving as Director of the Space Applications and Tactics Division. At the time, Space Command was working toward "operationalizing" the command's multiple space resources to effectively meet the needs of the war fighters. As a result, the command was looking for people with extensive operational experience such as Frank had.

A word about high-resolution radar

The SR-71 initially had an HRR (High Resolution Radar) system that, although remarkably effective for its time, became obsolete with advancements in satellite technology and imagery. By the mid-eighties, however, there was a growing demand from military and national command authorities for high-resolution, day/night, all-weather imagery which could only be partially satisfied with the available satellite technology. Consequently, the Air Force developed the ASARS-1 system (Advanced Synthetic Aperture Radar System I). It was first deployed on the SR-71 from Det 4 in the European Theater and, after some technical and operational "tweaking," turned out to be as good or better than high-resolution black and white photography. It was also highly accurate from as high as 85,000 feet. ASARS-1 was soon approved by the Joint Reconnaissance Center at the Pentagon for use in other theaters of operation worldwide. As an obvious corollary, the SR-71 mission-planning software needed to be updated to accommodate ASARS.

Life after the Air Force

Although he was not planning to retire from the Air Force at any particular time, Frank was unexpectedly offered a position by a Washington DC-based company to start up a new business unit in Colorado Springs. As a result, Frank retired from the Pentagon as a Colonel in 1993, where he had been Chief of the Space and Combat Integration Division. The Colorado company wanted to leverage the growing space-based opportunities in Colorado Springs by providing high-end graphics and analysis for the space industry. Frank served as General Manager of their Space Applications Division.

Frank retired from the defense contracting industry in 1999 and served as Executive Director of a local nonprofit agency helping homeless families with children. He finally fully retired in 2012.

Prior to that, Frank met his wife-to-be, another Air Force Officer, in Colorado Springs in 1990 during his assignment to Air Force Space Command. At the time of this writing, they have been married for 23 years.

Photos, courtesy of Frank Stampf

Gil and Frank

Frank and Gil, next sortie

Frank suiting up

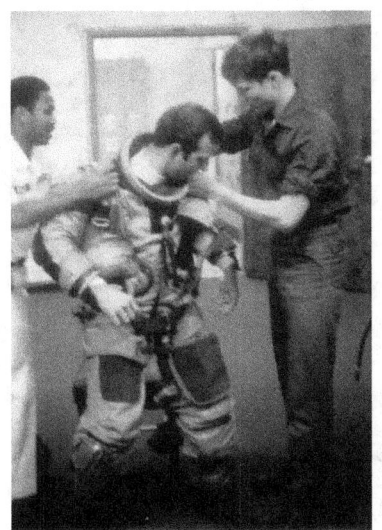

Frank – final suit check

Ian Campbell's Photograph Gallery (Ian's quotes)

"I had no idea in 1987 or 1992 that these materials had survived; I just had to hope. You only have to look at the hundreds of personal and official photos from Groom Lake (until 2015 it was a denied, non-existent facility) of aircraft, buildings, personnel, birthdays, laying concrete, and 100, 500, and 1,000 hours of flying – images we thought not only didn't survive, but also were never even allowed!"

Aircraft 929 undergoes maintenance in the background. This is the only picture of 929.

Some great historical shots

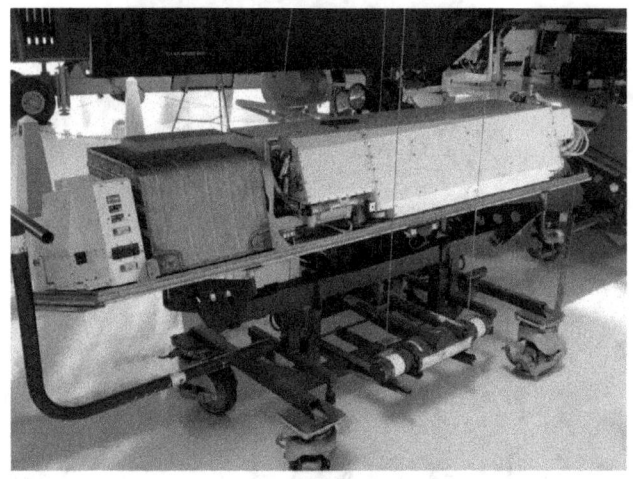

Ian Campbell's personal diagram of the first SR-71A flight

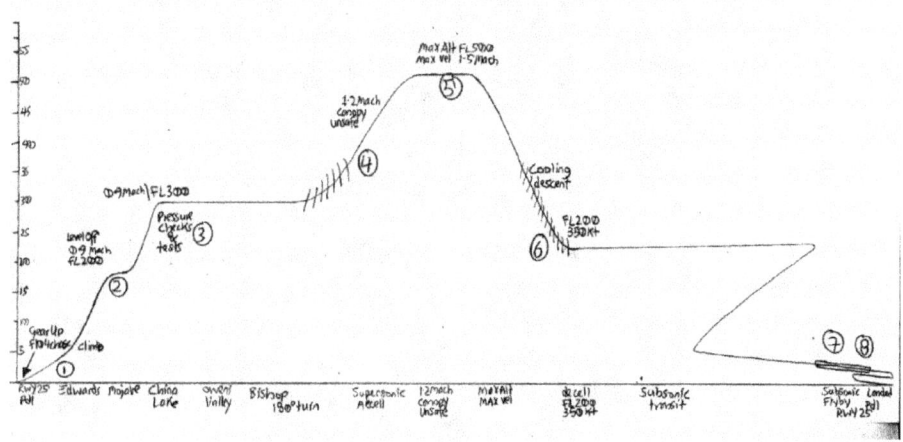

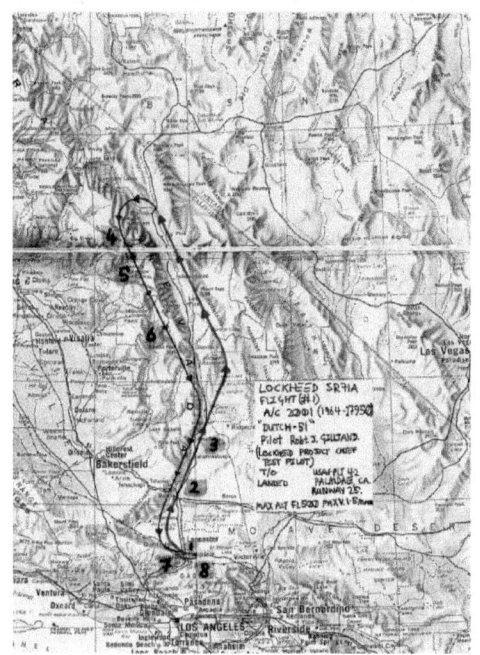

Later mission tracks

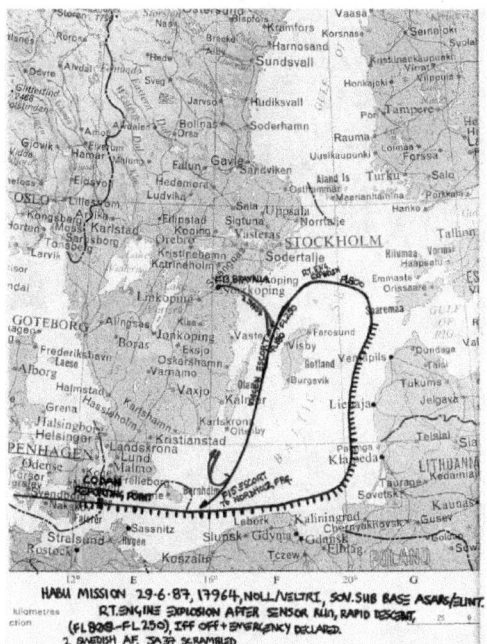

HABU MISSION 29.6.87, 17964, NOLL/VELTRI, SOV. SUB BASE ASW/ELINT. RT. ENGINE EXPLOSION AFTER SENSOR RUN, RAPID DESCENT (FL800-FL250), IFF OFF + EMERGENCY DECLARED. 2. SWEDISH AF. JA37 SCRAMBLED

24.5.77 '958: CARPENTER/MURPHY MLD-ARCP1- E.GER BORDER RUN- WILHEMSHAVEN- KAISERSLAUTEN- STUTTGART- SLAR, ELINT: 180° RUN AND MÜNCHEN TO NÜRNBERG, + BORDER.
31.5.77 '958: ALISON/VIDA MLD- RTB. RESAY 39.

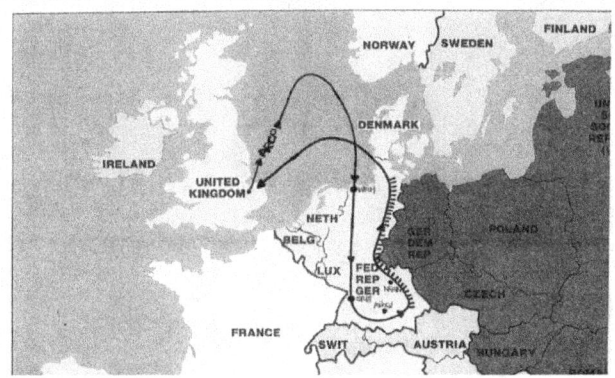

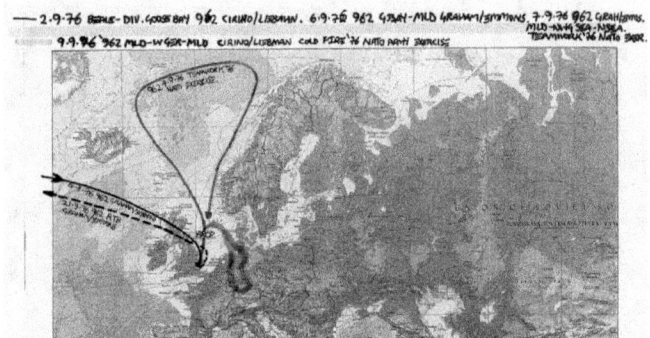

Mission Illustrations and Diagrams

By Ian, with the following comments: "I drew these 20-25 years ago, but they would not only work as mission illustrations, but also for illustrating what and where Mission Software integrated with Sensor packages, Autopilot, etc.

"Essentially, these were a simplification or extension of diagrams from either the Flight Manual, with climb and descent angles, or magazines and books describing a mission. What I added to the mission diagrams was more specific information than 1980's magazines had before such information was declassified: times, speeds, weights, etc., so that a mission could be laid out graphically for anyone to understand, especially the aerodynamic role of Dipsy-Doodle, Constant Cruise-Climb, Sensor Runs, Cooling Orbits, etc.

"A-11 to A-12 was my own development of a simple sketch in Squad-Sig 1055 – Blackbird in Action – from 1982, showing the difference chines made".

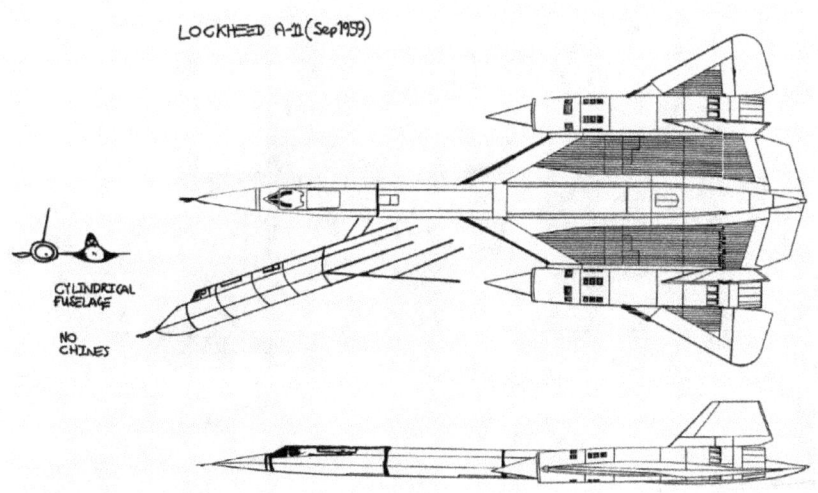

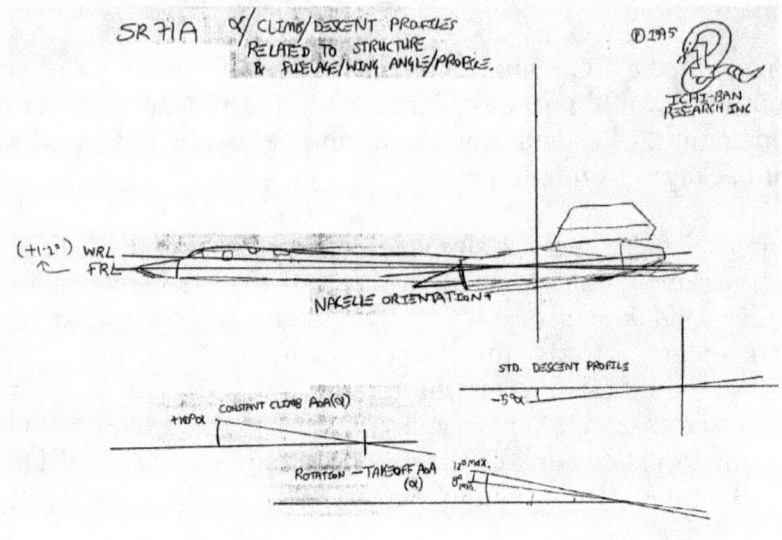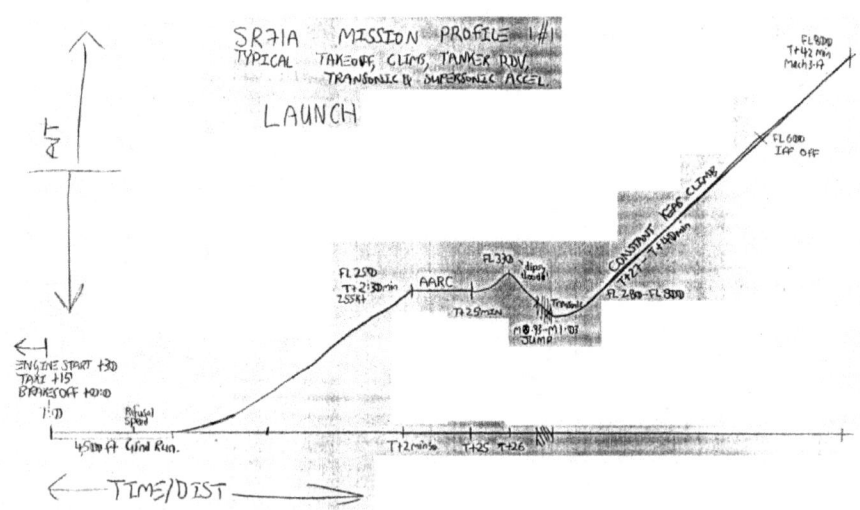

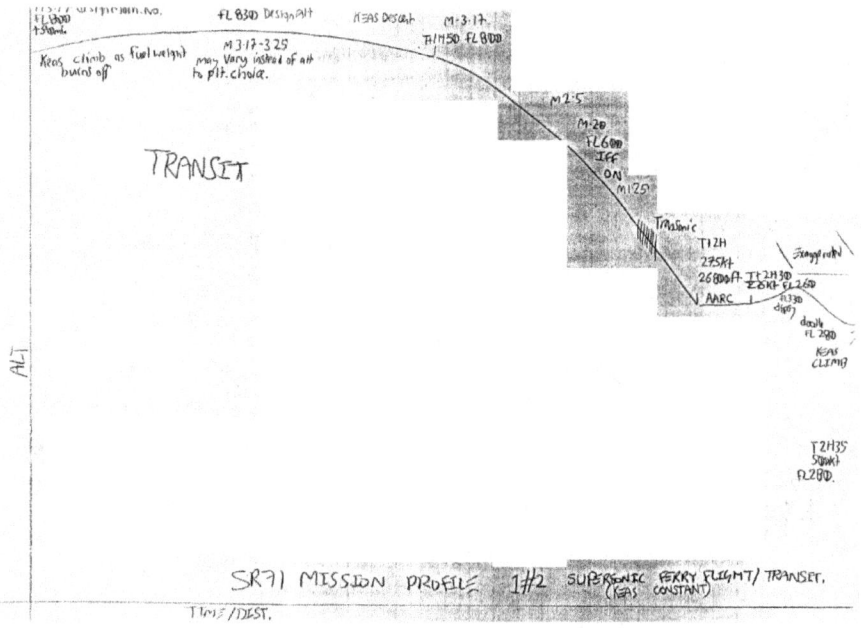

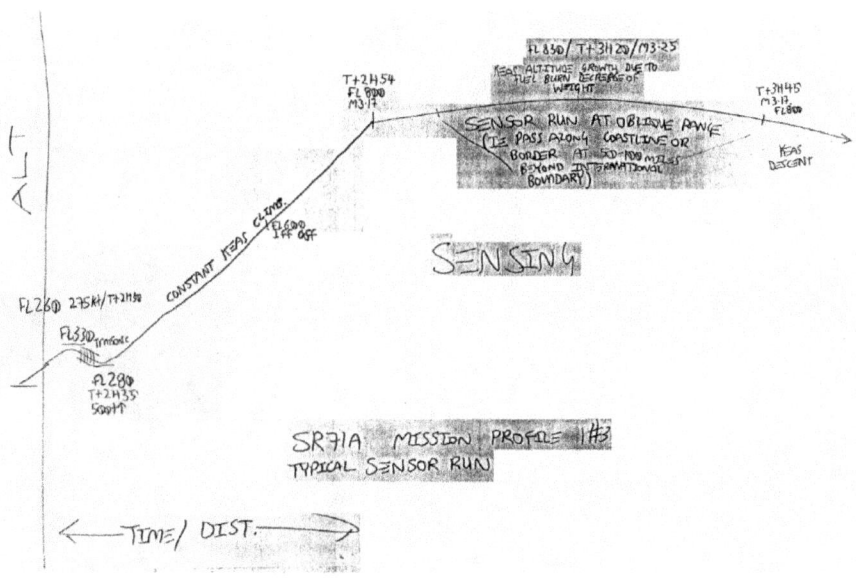

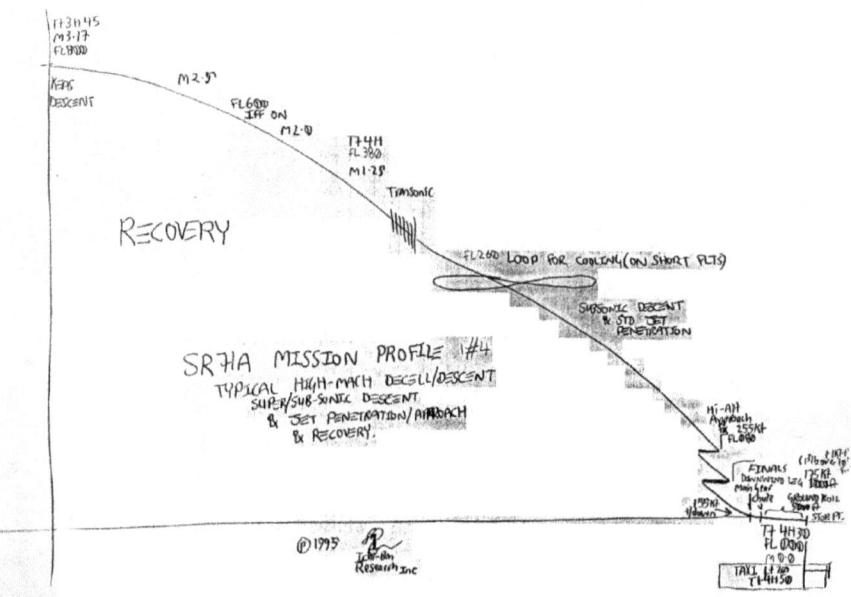

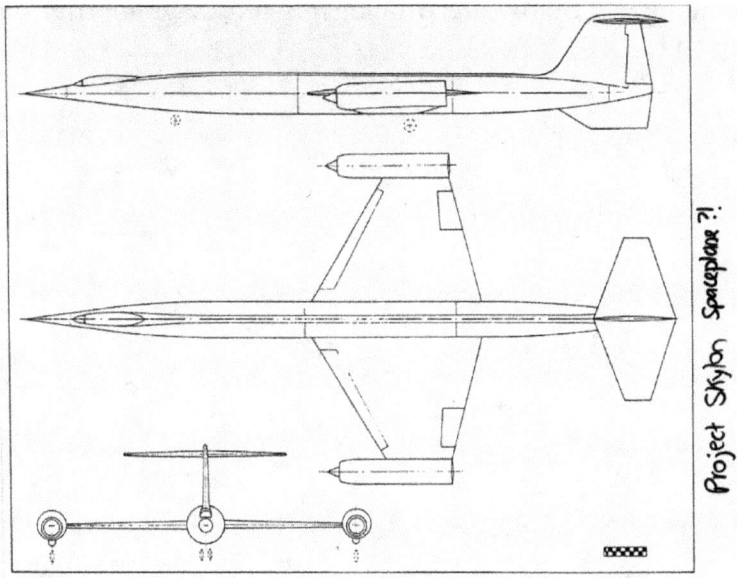

Lockheed CL-400 project

Lockheed CL-400

To meet the requirements for a U-2 successor, in early 1956 Kelly Johnson submitted to the USAF a proposal for a hydrogen-fuelled, high-altitude supersonic reconnaissance aircraft. An initial contract for two CL-400 prototypes was awarded in April 1956 and was followed soon after by an order for six more aircraft. Powered by two 9,500-lb (4,309-kg) thrust Pratt & Whitney 304-2 hydrogen-fuelled engines mounted at the tips of its thin trapezoidal wings, the CL-400 was a two-seat aircraft with span of 83 ft 9 in (25·53 m), length of 164 ft 10 in (50·24 m) and take-off weight of 69,955 lb (31,731 kg). Cruising at Mach 2·5 at between 95,000 and 100,000 ft (28,950 and 30,480 m), it was to reach targets within 1,100 naut miles (2,035 km) in fifty minutes. Engine development proceeded smoothly and manufacture was begun. However, Kelly Johnson and the Air Force were not satisfied with its short range. As 36,000 US gal (136,275 litres) of liquid hydrogen were already to be carried in three fuselage tanks, there appeared no feasible way to stretch range by increasing fuel capacity. Accordingly, the project was terminated in October 1957 and the almost completed prototypes were scrapped.

Ian's comment: "Below are official mission diagrams from a SAC briefing slide."

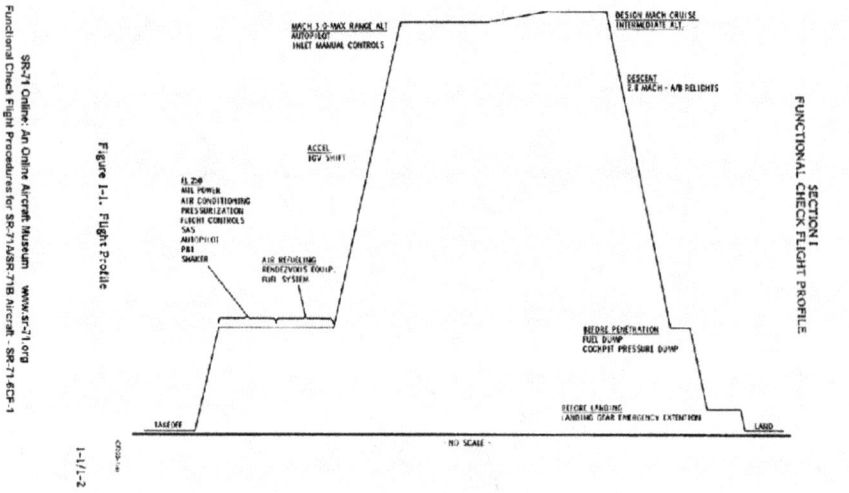

Fuel leaking under tank pressurization

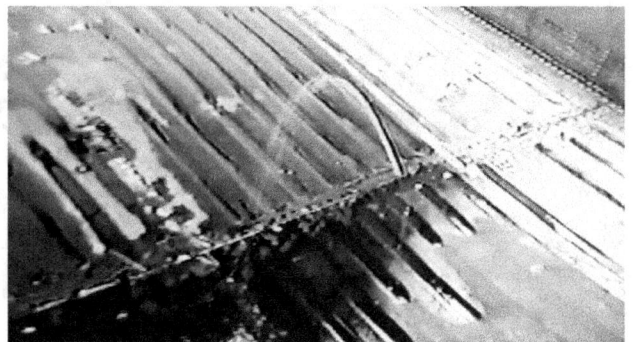

Finally, some great high-altitude photographs

APPENDIX A – A comprehensive crew list

Category	Rank	Last Name	FirstName/MI/Nickname	Date
Pilot	ADP	Schalk	Louis W. Jr.	26-Apr-62
VIP	ADP	Johnson	C.L. "Kelly"	1962
Pilot	ADP	Park	Bill	May-62
Pilot	ADP	Eastham	Jim	3-Feb-63
Pilot	ADP	Gilliland	Bob	5-Jan-63
Pilot	Civ	Skliar	William L.	4-Feb-63
Pilot	Civ	Collins	Kenneth S.	6-Feb-63
Pilot	Civ	Ray	Walter L.	10-Feb-63
Pilot	Civ	Sullivan	Dennis B.	14-Apr-63
Pilot	Civ	Walter	Alonzo J.	Feb-63
Eng	ADP	Edgar	Larry	Feb-64
Eng	ADP	Stockham	Hank	Feb-64
Eng	ADP	Larsen	Torrey	Feb-64
Pilot	ADP	Greenamyer	Darryl	25-Nov-64
FCO	Hughes	Byland	Tony	3-Dec-64
FCO	Hughes	Scalise	Ray	21-Dec-64
Eng	ADP	Miller	R. L. "Dick"	16-Feb-65
Pilot	ADP	Peterson	Art	20-Feb-65
FCO	Hughes	Parsons	George	24-Feb-65
Pilot	ADP	Weaver	Bill	18-Mar-65
RSO	ADP	Zwayer	Jim	29-Mar-65
FCO	Hughes	Archer	John	14-Apr-65
Eng	ADP	Beswick	Keith	1-Feb-63

RSO	ADP	Andre	George	16-Jun-65
RSO	ADP	Belgeau	Steven A.	17-Jun-65
RSO	ADP	Torick	Ray	17-Jun-65
Eng	ADP	Bohanan	Larry	18-Jun-65
Eng	ADP	Fulkerson	Glen	14-Dec-65
Pilot	Civ	Layton	Ronald "Jack"	15-Apr-65
Pilot	Civ	Vojvodich	Mele	3-Mar-65
Pilot	Civ	Weeks	Jack C.	21-Sep-65
RSO	ADP	Moeller	Kenneth E.	6-Oct-65
VIP	ADP	Daniell	Rus	7-Jul-66
Pilot	Civ	Murray	Francis J.	25-Jun-67
Pilot	Civ	Scott	Russell J.	1-Apr-67
Pilot	Col	Stephens	Robert L. "Fox"	27-Jan-65
FCO	LtCol	Andre	Daniel	14-Jan-65
FCO	Maj	Warner	Noel T.	27-Jan-65
FCO	Capt	Cooney	James P.	20-Feb-65
Pilot	Maj	Daniel	Walter F.	14-Apr-65
FCO	Capt	Ursini	Sammel M.	22-Apr-65
RSO	Maj	Hurley	Kenneth D.	28-Apr-65
Pilot	Maj	Bock	Charles C.	3-Jun-65
Pilot	Col	Henderson	Vern	20-Jul-65
Pilot	LtCol	Beezley	Jacques G.	2-Aug-65
Pilot	Col	Templeton	H. A.	6-Aug-65
Pilot	Maj	Evenson	Mervin L.	2-Sep-65
Pilot	Maj	Hichew	Allen L.	18-Nov-65
RSO	Capt	Schmittou	Tom W.	18-Nov-65
Pilot	Maj	Sowers	Robert G. "Gray"	30-Nov-65
RSO	Capt	Sheffield	Richard E. "Butch"	30-Nov-65
RSO	Capt	Mallozzi	Cosimo B. "Coz"	2-Dec-65
Pilot	LtCol	Haupt	Ray	6-Dec-65
RSO	LtCol	Braeden	Cecil H.	19-Apr-66
RSO	Capt	Fagg	James W.	22-Apr-66
RSO	Maj	Payne	William R.	26-Apr-66

Pilot	Col	Bennett	Frenchy Dumont	16-May-66
VIP	BrGen	Manson	Hugh B.	May-66
Pilot	LtCol	Rogers	Joe	19-May-66
Staff	LtCol	Richardson	Ralph	26-May-66
Pilot	Col	Holbury	Robert J.	2-Jul-65
Pilot	Col	Slater	Hugh C. "Slip"	30-Jun-65
Pilot	Maj	Burgeson	Harold E.	24-Jun-65
Pilot	LtCol	Barrett	Burton S.	18-Feb-66
Pilot	LtCol	Perkim	Roland L.	19-Apr-66
RSO	LtCol	Drake	Norman S.	7-Jun-66
Pilot	Maj	Bowles	Ben	13-Jun-66
RSO	LtCol	Peterson	Harold C.	24-Jun-66
Pilot	Maj	Campbell	William J.	13-Jul-66
RSO	Capt	Pennington	Albert N.	13-Jul-66
Pilot	Col	Nelson	Douglas T.	28-Jul-66
RSO	LtCol	Lewis	Russel L.	28-Jul-66
Pilot	Maj	Halloran	Patrick J.	4-Aug-66
RSO	Capt	Jarvis	Mort J.	4-Aug-66
RSO	Capt	Dempster	David P.	12-Aug-66
Pilot	Maj	Brown	Buddy L.	18-Aug-66
RSO	Capt	Jensen	David J.	18-Aug-66
Pilot	Capt	Collins	Charles W. "Pete"	10-Aug-66
RSO	Capt	Seagroves	Jean C. "Conrad"	10-Aug-66
Pilot	Maj	Storrie	John H.	17-Aug-66
Pilot	Maj	Kennon	Jack	27-May-66
Pilot	Capt	Shelton	Franklin D. "Dale"	9-Sep-66
RSO	Capt	Boggess	Lawrence L.	9-Sep-66
Pilot	LtCol	Skliar	William L.	24-Aug-66
Pilot	Maj	O'Malley	Jerome F.	20-Oct-66
RSO	Capt	Payne	Edward D.	20-Oct-66
Pilot	Maj	Walbrecht	Donald A.	10-Nov-66
RSO	Capt	Loignon	Phillip G.	10-Nov-66
Pilot	Capt	Boone	Earle M.	22-Nov-66
RSO	Capt	Vick	Dewain C.	22-Nov-66
Staff	Col	Weller	Russell K.	6-Jan-67

Pilot	Maj	Watkins	James L.	17-Feb-67
Pilot	Maj	Bevacqua	Anthony P.	23-Feb-67
Staff	Col	Hayes	William P.	27-Feb-67
RSO	Capt	Roetcisoender	Robert J.	15-Mar-67
RSO	Maj	Keller	William C.	17-Mar-67
RSO	Maj	Mathers	Donald E.	24-Mar-67
Pilot	Capt	Spencer	Robert C.	7-Apr-67
Pilot	Col	Confer	Harold E.	11-Apr-67
RSO	Capt	Shoemaker	Clyde L.	10-Apr-67
Pilot	Maj	McCallom	Brian K.	5-May-67
	LtCol	Schever	James C.	25-Mar-66
RSO	Maj	Crew	Jerald L.	15-May-67
Pilot	Maj	Devall	Larry S.	9-May-67
RSO	Capt	Branham	Ruel K.	18-May-67
RSO	Maj	Casey	Thomas A.	25-May-67
Pilot	Maj	St.Martin	Roy L.	29-May-67
RSO	Capt	Locke	Robert M.	8-Jun-67
Pilot	Maj	Bull	George M.	19-Jun-67
RSO	Capt	Carnochan	John A.F.	23-Jun-67
Staff	Col	Boynton	John B.	29-Jun-67
Pilot	Maj	Powell	Robert M.	5-Jul-67
RSO	Maj	McNeer	Charles J. "Red"	17-Jul-67
Pilot	Maj	Daubs	Charles E.	24-Jul-67
ENG	Maj	Lusby	William A. "Art"	15-Aug-67
ENG	AF/Civ	Sudderth	Robert W.	18-Aug-67
RSO	Capt	Kendrick	William J.	30-Aug-67
Staff	Col	Minter	Charles F. Sr.	6-Sep-67
ENG	AF/Civ	Abrams	Richard	12-Sep-67
Pilot	Capt	Campbell	Bobby L.	31-Aug-67
VIP	Gen	Holloway	Bruce (CINC)	4-Oct-67
VIP	LtGen	Agan	Arthur C.	12-Oct-67
VIP	LtGen	Martin	William K.	20-Oct-67
Pilot	Maj	Kardong	Gabriel A. "Abe"	17-Oct-67
VIP	LtGen	Catton	Jack J. (15AF/CC)	9-Oct-67

RSO	Capt	Heidlebaugh	Gary	20-Oct-67
RSO	Maj	Kraus	Jon P.	27-Oct-67
Pilot	Maj	Maier	Lothar J.	17-Nov-67
Pilot	Maj	Lawson	William E. III	7-Dec-67
RSO	Maj	Kogler	James A.	15-Dec-67
RSO	Maj	Coleman	Gary L.	Dec-67
RSO	Maj	Martinez	Gilbert	22-Dec-67
RSO	LtCol	Chapman	Harold E.	16-Jan-68
Pilot	Capt	Fruehauf	David E.	9-Feb-68
VIP	Col	Bellis	Benjamin N.	1968
RSO	Capt	Payne	Allen R.	27-Mar-68
Pilot	Maj	Hudson	James W.	15-Mar-68
RSO	Capt	Budzinski	Norbert L.	15-Apr-68
RSO	Maj	Byrnes	Donn A.	18-Apr-68
RSO	Maj	Hartman	Bruce	23-Apr-68
RSO	Maj	Weaver	R.W. Jr.	27-May-68
Pilot	LtCol	Hain	Harlon A.	27-May-68
RSO	Maj	Rhude	Don	23-May-68
Pilot	Maj	Gerard	Richard C.	21-Jun-68
Pilot	Maj	Estes	Thomas B.	24-Jul-68
RSO	Maj	McLean	Charles G.	19-Jul-68
RSO	Maj	Moeller	Ted G.	30-Sep-68
Pilot	LtCol	Collins	Kenneth S.	17-Feb-65
Pilot	Maj	Shelton	James H. Jr.	4-Aug-68
Pilot	Capt	Allender	Reverdy J.	9-Dec-68
	LtCol	Trost	Fred	16-Dec-68
Pilot	LtCol	Sullivan	Dennis B.	14-Apr-65
Pilot	LtCol	Vojvodich	Mele	3-Mar-65
Pilot	Capt	Wilcox	Bruce E.	4-Mar-69
Pilot	Maj	Pyne	E.L. (EAFB)	10-Mar-69
VIP	Senator	Goldwater	Barry	2-Apr-69
Staff	LtCol	Layton	Ronald J. "Jack"	15-Apr-65
Staff	Col	Anderson	James E.	4-Jun-69
VIP	BrGen	Slay	Alton D.	7-Aug-69
VIP	Col	Payne	William R.	18-Sep-69
Pilot	Maj	Cobb	Darrel W.	10-Nov-69

RSO	Capt	Gantt	Myron L.	30-Dec-69
Pilot	Mr/NASA	Fulton	Fitzhugh L. Jr.	4-Mar-70
Pilot	Maj	Pugh	Thomas S.	11-Mar-70
Pilot	Mr/NASA	Mallick	Donald L.	10-Apr-70
RSO	Maj	Rice	Ronnie C.	15-Apr-70
Pilot	Capt	Bush	Dennis K.	21-Apr-70
Pilot	Capt	Hertzog	Randolph B.	31-Aug-70
Pilot	Col	Sullivan	D.B.	18-Sep-70
RSO	Maj	Selberg	Ronald L.	23-Oct-70
Staff	LtCol	Owen	Roy W. Jr.	9-Oct-70
RSO	Maj	Curtis	Billy A.	17-Nov-70
Pilot	Maj	Sewell	George H. Jr.	13-Dec-70
RSO	Capt	Blackwell	Reginald T.	2-Jan-71
VIP	LtGen	Carlton	Paul K. (15AF/CC)	2-Mar-71
RSO	Capt	Morgan	George T.	3-Mar-71
Pilot	Capt	Cunningham	Robert J.	19-Feb-71
RSO	Mr/NASA	Horton	Victor W.	22-Mar-71
VIP	Col	Sprinkle	Ross	23-Apr-71
Pilot	Capt	Judkin	Monty T.	5-May-71
VIP	BrGen	Lukeman	Robert F.	13-Nov-71
Pilot	Maj	Gunther	Caroll D.	24-Nov-71
RSO	Mr/NASA	Young	William R.	30-Nov-71
RSO	Capt	Allocca	Thomas R.	9-Dec-71
Staff	BrGen	Harris	Edgar S. Jr.	4-Dec-71
VIP	Cong.	Price	Robert	12-Feb-72
Pilot	Capt	Haller	Carl A.	25-Feb-72
VIP	MajGen	Felices	Salvador E.	24-Mar-72
cancelled				
Pilot	Maj	Bledsoe	A.H. "Pat"	7-Jun-72
RSO	Maj	Rogers	Cletius C.	27-Jun-72
Pilot	Maj	Sullivan	James V.	23-Jul-72
RSO	Capt	Widdifield	Noel E.	11-Aug-72

RSO	Capt	Fuller	John T.	9-Mar-72
Pilot	Capt	Ransom	Leland B. III	14-Jul-72
RSO	Maj	Gersten	Mark H.	1-Aug-72
Pilot	Capt	Adams	Harold B. "Buck"	18-Oct-72
RSO	Capt	Machorek	William C.	1-Nov-72
Pilot	Capt	Helt	Robert C.	3-Jan-73
RSO	Capt	Elliott	Larry A.	17-Jan-73
Pilot	Capt	Joersz	Eldon W.	30-Jan-73
Pilot	Capt	Wilson	James F.	26-Feb-73
VIP	MajGen	McLaughlin	George W.	
RSO	Capt	Douglass	Bruce S.	27-Mar-73
VIP	Dr./Sec of AF		John L.	6-Sep-73
VIP	LtGen	Pitts	William F. (15AF/CC)	6-Jul-73
VIP	BrGen	Pittman	Don D. (14AD/CC)	20-Sep-73
Pilot	Capt	Rosenberg	Maury	6-Nov-73
RSO	Capt	Bulloch	Donald C.	21-Nov-73
VIP	LtCol	Rupard	Hal	23-Jan-74
VIP	LtCol	Reed	Jackie G.	23-Jan-74
Pilot	LtCol/AFLC		Tom	23-Jan-74
Staff	LtCol	Samay	Raphael S.	5-Mar-74
Pilot	Capt	Kinego	Joseph C.	22-Apr-74
RSO	Capt	Jacks	Roger L.	8-May-74
VIP	MajGen	Sitton	Ray B.	21-Jun-74
VIP	Honorable	Currie	Malcolm R.	1-Jul-74
VIP	Honorable	Laberge	Walter B.	1-Jul-74
Pilot	Capt	Cirino	Alan B.	26-Aug-74
RSO	Capt	Liebman	Bruce L.	17-Sep-74
Pilot	Capt	Murphy	Justin J. "Jay"	24-Oct-74
VIP	BrGen	Melton	Albert L.	
RSO	Maj	Billingsley	John A.	27-Dec-74
VIP	LtGen	Keck	James M. (VCINC)	13-Dec-74
VIP	MajGen	Ellis	Billy J.	12-Feb-75
Pilot	Capt	Graham	Richard H.	27-Feb-75
RSO	Capt	Emmons	Donald R.	17-Mar-75

VIP	MajGen	Gavin	Herbert J.	7-Aug-75
Pilot	Capt	Alison	Thomas M.	28-May-75
Pilot	Maj/AFLC	Riedenauer	Robert L.	5-Jun-75
RSO	Maj/AFLC	Frazier	William J.	4-Oct-75
RSO	Maj	Vida	Joseph T.	18-Jun-75
VIP	Col	Dethlefsen	Merlyn H.	24-Jun-75
VIP	LtGen	Shotts	Bryan M. (15AF/CC)	19-Sep-75
VIP	LtGen	Anderson	Andrew B. Jr.	26-Sep-75
VIP	MajGen	Burkhart	John W.	24-Oct-75
Staff	Col	Beckel	Robert D.	11-Nov-75
VIP	MajGen	Moore	Otis	10-Dec-75
Staff	Col	Bower	Richard J.	12-Dec-75
Pilot	Capt	Gilmore	William G.	30-Dec-75
Pilot	Maj	Crowder	Robert W.	1-Apr-76
RSO	Capt	Morgan	John G.	21-Apr-76
VIP	Gen	Dougherty	Russell E. (CINC)	24-Jun-76
Pilot	Capt	Carpenter	Adelbert W. "Buz"	9-Jul-76
VIP	LtGen	Roberts	John W.	29-Jul-76
RSO	Capt	Murphy	John E.	29-Jul-76
VIP	BrGen	Cooke	Gerald E.	1976
VIP	Mr.	Plummer	James W.	1976
RSO	Capt	Keller	William C.	15-Sep-76
VIP	LtGen	Hails	Robert E.	4-Nov-76
Staff	Col	Kidder	Lyman M.	12-Nov-76
VIP	Honorable	Connor	James B.	1976
Pilot	Maj	Veth	John J. "Jack"	20-Dec-76
Pilot	Maj	Groninger	William G.	24-Mar-77
RSO	Capt	Sober	Charles T. Jr.	18-Apr-77
Pilot	Maj	Thomas	Bredette	1-Sep-77
VIP	BrGen	Brown	Bill V. (14AD/CC)	1977
RSO	Capt	Reid	Jay	22-Sep-77

VIP	MajGen	Peck	Earl G.	1977
Staff	Col	Fenimore	John W.	13-Oct-77
Pilot	Capt	Keck	Thomas J.	31-Oct-77
RSO	Capt	Shaw	Timothy J.	22-Nov-77
Pilot	Mr/NASA	Manke	John	1-Dec-77
Pilot	Mr/NASA	Dana	Bill	1-Dec-77
Pilot	Mr/NASA	Krier	Gary E.	9-Dec-77
Pilot	Mr/NASA	Enevoldson	Einar K.	13-Dec-77
Pilot	Mr/NASA	McMurtry	Thomas C.	14-Dec-77
Pilot	Maj	Peters	David M.	30-Dec-77
RSO	Capt	Stockton	Michael L.	13-Feb-78
VIP	Col	Shadburn	Ted H.	15-May-78
RSO	Capt	Bethart	Edgar J. Jr.	16-Jun-78
Pilot	Maj	Shelton	Lee M. III	18-Jul-78
Staff	Col	Terry	Andrew G.	1-Aug-78
Staff	Col	Young	David G.	29-Aug-78
VIP	LtGen	Mathis	Robert C.	1978
RSO	Maj	MacKean	Barry C.	8-Sep-78
Pilot	LtCol AFLC	Jewett	Calvin F.	7-Nov-78
Pilot cancelled	Capt	Young	Richard A.	21-Dec-78
RSO	Capt	Szczepanik	Russell L.	24-Jan-79
VIP	Reverend	Hesburgh	Theodore M.	28-Feb-79
VIP	LtGen	Mullins	James P. (15AF/CC)	29-Mar-79
Pilot	Mr/NASA	Ishmael	Stephen D.	29-Mar-79
Pilot	Mr/NASA	Swann	Michael R.	29-Mar-79
VIP	BrGen	Brashear	John A. (14AD/CC)	23-Apr-79
Pilot	Capt	Augustin	Calvin J.	17-May-79
RSO	Maj	Kelly	Frank K.	25-Jun-79
VIP	Col	Stanton	Joseph S.	24-Jun-79

VIP	Senator	Cannon	Howard W.	18-Jan-80	
Pilot	Capt	Bertelson	Gilbert M.	27-Dec-80	
RSO	Capt	Stampf	Frank W.	29-Jan-80	
VIP	Mr.	Fishburne	Frank A.	12-Mar-80	
Pilot	Capt	Judson	Richard W.	24-Mar-80	
cancelled					
VIP	LtGen	Leavitt	Lloyd R. Jr.	8-May-80	
Pilot	Maj	Cunningham	Nevin N.	22-May-80	
RSO	Capt	Quist	Gene R.	16-Jun-80	
cancelled					
Pilot	Capt	Berg	Dennis R.	26-Aug-80	
RSO	Maj/AFLC	Flanagan	Bill	10-Sep-80	
VIP	Mr.	Hartz	Jim	28-Aug-80	
RSO	Capt	McKim	E. D.	5-Nov-80	
Pilot	Maj	Glasser	Gerald T.	19-Nov-80	
cancelled					
RSO	Capt	Hornbaker	David M.	18-Dec-80	
VIP	Mr.	Ropelewski	Robert R.	1981	
Pilot	Capt	McCrary	Richard S.	8-Apr-81	
RSO	Capt	Lawrence	David A.	28-Apr-81	
Pilot	Maj	Smith	Bernard J.	11-Jun-81	
VIP	MajGen	Chain	John T. Jr.	15-Jul-81	
RSO	Maj	Whalen	Dennis W.	16-Jul-81	
VIP	BrGen	Hatch	Monroe W. (14AD/CC)	5-Nov-81	
Pilot	Capt	Luloff	Gary I.	3-Nov-81	
VIP	Gen	Davis	Bennie L. (CINC)	11-Dec-81	
RSO	Capt	Coats	Robert L.	3-Dec-81	
VIP	LtGen	Murphy	John J. (15AF/CC)	8-Jan-82	
VIP	Cong.	Dornan	Robert K.	8-Feb-82	
Pilot	Maj	Dyer	Leslie R.	5-Feb-82	
RSO	Capt	Greenwood	Daniel	9-Mar-82	
VIP	LtGen	Miller	George D. (VCINC)	19-Mar-82	

VIP	MajGen	Buckman	Louis C. (SAC/DO)	10-Apr-82
Staff	Col	Liss	Lonnie L.	16-Apr-82
Pilot	Capt	Burk	William	21-Apr-82
Pilot	Maj/AFLC	Tilden	Thomas V.	12-May-82
VIP	Gen	Allen	Lew Jr. (CS/Air Force)	4-Jun-82
RSO	Capt	Henichek	Thomas J.	23-Jun-82
VIP	BrGen	Hocker	Jesse S. (14AD/CC)	19-Aug-82
Pilot	Capt	Jiggens	James M.	2-Sep-82
RSO	Capt	McCue	Joseph J.	24-Sep-82
VIP	Sec. of AF	Aldridge	Edward C. "Pete" Jr.	22-Sep-82
VIP	LtGen	Iosue	Andrew P.	5-Nov-82
Pilot	Maj	Behler	Robert F.	2-Nov-82
RSO	Capt	Tabor	Ronald D.	7-Dec-82
VIP	Cong.	Young	C.W. "Bill"	19-Feb-83
Pilot	Maj	Boudreaux III	Lionel P.	3-Feb-83
RSO	Capt	Newgreen	Walter F. "Terry"	7-Mar-83
VIP	LtGen	Skantze	Lawrence A.	19-Aug-83
Pilot	Maj	Matthews	Joseph E.	29-Jun-83
RSO	Capt	Ross	Edward W.	1-Jul-83
VIP	BrGen	Yeager	Charles E. "Chuck"	5-Aug-83
RSO	Capt	Osterheld	D. Curt	11-Aug-83
Staff	Col	Freese	George V.	19-Sep-83
Staff	Col	Pinsky	David H.	22-Sep-83
Pilot	Capt	Madison	Jack E.	7-Oct-83
VIP	LtGen	Light	James E. Jr. (15AF/CC)	23-Nov-83
RSO	Maj	Orcutt	William D.	9-Nov-83
Pilot	Capt	Yeilding	R. Edward	19-Dec-83

RSO	Capt	Lee	Stephan M.	18-Jan-84
Pilot	Maj	Shul	Brian	10-Jul-84
RSO	Maj	Watson	Walter L.	23-Aug-84
Pilot	Capt	Deal	Duane W.	12-Sep-84
RSO	Capt	Veltri	Thomas F.	5-Oct-84
Pilot	Maj	Noll	Duane M.	14-Dec-84
RSO	Capt	Morgan	Charles A.	14-Jan-85
VIP	Cong.	Stump	Bob	1-Apr-85
Pilot	Maj	Dyckman	William R. "Rod"	13-Mar-85
RSO	Maj	Bergam	Thomas E.	7-Apr-85
Staff	BrGen	Farrington	John (14AD/CC)	4-Jun-85
Pilot	Maj	Smith	Michael L.	11-Jun-85
RSO	Capt	Soifer	Douglas B.	12-Jul-85
Staff	Col	McConnell	Robert B.	25-Jul-85
Staff	MajGen	Davidson	Alexander K. (AF/XO)	23-Oct-85
RSO	Maj/AFLC	Soucy	Philip L.	1-Nov-85
VIP	Cong.	Byron	Beverly	18-Nov-85
Pilot	Maj	House	Dan E.	4-Dec-85
RSO	Capt	Bozek	Blair L.	22-Jan-86
VIP	Gen	Welch	Larry D. (CINC)	15-Jan-86
Pilot	Maj	Danielson	Thomas J.	21-Apr-86
VIP	Cong.	Badham	Robert E.	30-Apr-86
RSO	Maj	Gudmundson	Stanley J.	20-May-86
Pilot	Maj	Pappas	Terry D.	17-Jun-86
RSO	Capt	Manzi	John D.	14-Jul-86
Staff	Col	Savarda	James S.	9-Sep-86
VIP	MajGen	Johnson	Hansford (SAC/DO)	10-Oct-86
Pilot	Maj	McKendree	Warren C.	29-Sep-86
RSO	Maj	Shelhorse	Randy F.	12-Nov-86
VIP	Cong.	Hopkins	Larry J.	14-Nov-86
Pilot	Maj	Brown	Larry	23-Dec-86

RSO	Maj	Carter	Keith E.	20-Feb-87
Pilot	Capt	Grzebiniak	Steven	13-Mar-87
RSO	Capt	Greenwood	James F.	23-Apr-87
Pilot	Maj	McCleary	Thomas R.	3-Aug-87
RSO	Capt	Vardaman	Hunter W.	27-Aug-87
VIP	LtGen	Peek	Kenneth L. Jr. (VCINC)	15-Oct-87
Staff	BrGen	Estes	Howell M. III (14AD/C)	5-Nov-87
Staff	Col	Schreiber	Donald R.	7-Dec-87
Pilot	Capt	Crittenden	Gregory N.	21-Dec-87
RSO	Maj/AFLC	Fuhrman	Tom	21-Jan-88
RSO	Maj	Finan	Michael J.	28-Jan-88
Pilot	Maj	Watkins	Donald T.	16-Feb-88
RSO	Capt	Fowlkes	Robert E.	24-Mar-88
Pilot	Capt	Snyder	Ben	13-Apr-88
RSO	Capt	Shade	Briggs	19-Apr-88
VIP	LtGen	Burpee	Richard A. (15AF/CC)	5-Aug-88
Staff	BrGen	Keller	Kenneth F. (14AD/CC)	9-Sep-88
VIP	LtGen	Shuler	Ellie G. "Buck" Jr. (8AF/CC)	21-Dec-88
Staff	Col	Grimes	William (AFLC)	19-Jan-89
Pilot	Maj	Halsell	Jim	11-Apr-89
Staff	MajGen	Borling	John L.	20-Jul-89
Pilot	Mr/NASA	Smith	Rogers	14-Aug-91
FE	Ms/NASA	Bohn-Meyer	Marta	3-Oct-91
FE	Mr/NASA	Meyer	Robert R.	9-Oct-91
Staff	Mr/NASA	McMurtry	Tom	4-Jun-92
VIP		Barthelemy	Robert	26-Jun-92

VIP	Gen	Dailey	John R. (USMC)	25-Feb-94
Pilot	Mr/NASA	Schneider	Ed	18-Oct-94
Staff	Mr/NASA	Fullerton	C. Gordon	17-Nov-94
VIP	MajGen	Engel	Richard L.	14-Dec-95
Staff	Mr/NASA	Purifoy	Dana D.	7-Sep-96
Staff	Mr/NASA	Stucky	Mark	4-Mar-97
Staff	Mr/NASA	Knutson	Marty	17-Apr-97
Pilot	Maj	Garrison	Bert	24-Jun-97
RSO	Maj	Ochotorena	Dom	25-Jul-97
		Barnett	Larry	11-Oct-73
		Kelly	John	1963
		Ledford	Jack	1963
		Perkins	Cy	1963
Staff	Maj	Spruill	Dameron	8-May-69
VIP	Col	Uppstrom	Robert	7-Nov-79
		Young	David P.	1963

Conclusion

Hopefully, after reading this book you have a sense of the magnitude of our software accomplishment.

The Habu / Blackbird SR-71 flew for twenty-three years using our original software. Yes, it was updated for new sensors and a later version of FORTRAN, but the base logic remained.

Compared with the software tools and computer hardware available today, our Herculean effort was done with just "bearskins and knives."

And many thanks to

- Our talented proof readers: Linda Cooper, Jack Hirschfield and Jim Peragallo.
- For those interviewed, who took the time to recall their stories
- And, finally, to the brave pilots and RSOs who made the SR-71 a legend in its time

www.ingramcontent.com/pod-product-compliance
Lightning Source LLC
Chambersburg PA
CBHW052251220526
45471CB00001B/288